ALTERNATIVE USES FOR AGRICULTURAL SURPLUSES

Proceedings of the Seminar on 'Research and the Problems of Agricultural Surpluses in Europe', organised by the Directorate-General for Agriculture (DG VI) and the Directorate-General for Science, Research and Development (DG XII) of the Commission of the European Communities, and held in Brussels, Belgium, on 25–27 June 1985.

Organising Committee:

G. Fauconneau (*France*), W. F. Raymond (*United Kingdom*), D. Schliephake (*Federal Republic of Germany*), J. Gillot, J. Dehandtschutter and Christa Haebler (*Commission of the European Communities, DG VI*), P. Larvor and P. Reiniger (*Commission of the European Communities, DG XII*).

ALTERNATIVE USES FOR
AGRICULTURAL SURPLUSES

Edited by

W. F. RAYMOND

Consultant, United Kingdom

and

P. LARVOR

Commission of the European Communities, Brussels

ELSEVIER APPLIED SCIENCE
LONDON and NEW YORK

ELSEVIER APPLIED SCIENCE PUBLISHERS LTD
Crown House, Linton Road, Barking, Essex IG11 8JU, England

Sole Distributor in the USA and Canada
ELSEVIER SCIENCE PUBLISHING CO., INC.
52 Vanderbilt Avenue, New York, NY 10017, USA

WITH 21 TABLES AND 86 ILLUSTRATIONS

© ECSC, EEC, EAEC, BRUSSELS AND LUXEMBOURG, 1986

British Library Cataloguing in Publication Data

Seminar on Research and the Problems of Agricultural
Surpluses in Europe *(1985: Brussels)*
Alternative uses for agricultural surpluses.
1. Agricultural processing—European Economic
Community countries 2. Surplus agricultural
commodities—European Economic Community countries
I. Title II. Raymond, W. F. III. Larvor, P.
IV. Commission of the European Communities.
Directorate-General for Agriculture V. Commission
of the European Communities. *Directorate-General
for Science, Research and Development*
631 S698

ISBN 1-85166-084-4

Library of Congress CIP data applied for

Publication arrangements by Commission of the European Communities, Directorate-General Telecommunications, Information Industries and Innovation, Luxembourg

EUR 10603 EN

LEGAL NOTICE

Neither the Commission of the European Communities nor any person acting on behalf of the Commission is responsible for the use which might be made of the following information.

Printed in Great Britain by Galliard (Printers) Ltd, Great Yarmouth

v

CONTENTS

SURPLUSES AND THE COMMON AGRICULTURAL POLICY

R. MILDON
CEC, DG VI (Agriculture)

Summary

The European Community has a substantial overall balance of payments deficit for agricultural and food products. Nevertheless the Community produces, for certain commodities, more than its own domestic and export markets require. When production exceeds demand, stocks are usually created. Generally if a structural surplus develops, stocks grow to excessive levels; the classic example of a structural surplus is indicated by the 'butter mountain'. However the existence of stocks is not in itself confirmation that a structural surplus exists, while structural surpluses exist for other commodities where stocks are at lower levels - for example, wine. The phenomenon of surpluses manifests itself in different ways specific to the commodity in question and this report covers the main agricultural market organisations where surpluses occur or might occur.

INTRODUCTION

Those of us engaged in the management and development of the Common Agricultural Policy are convinced that there is a problem of agricultural surpluses, our difficulty has been and remains one of formulating that problem in such a way that everyone agrees on the solution. The difficulty for the Commission of the European Communities is to formulate the problem of surpluses in such a way that everyone involved in the decision making process can accept it and in such a way that the solutions are compellingly self-evident.

The Commission is at present actively engaged in preparing a 'Green paper' which will be entitled "Perspectives for the Common Agricultural Policy", a consultative document. Progress will be made when the situation develops where those responsible for the Common Agricultural Policy agree on a coherent approach to the medium and long term development of this common policy. As recent events in the Council of Ministers have demonstrated, a significant minority of interested parties do not agree on the short term adaptation of the agricultural policy. Their objections are not only that a modest cut in one or two institutional prices in nominal terms are unacceptable - because of the possible deterioration in farming incomes - but also because they do not agree with the way in which the Commission has formulated the problem.

Let us first try to pin down what we mean by an agricultural surplus. This in itself is no easy task and although one may have recourse to a dictionary, the static definitions which it supplies are incomplete when dealing with the real world which is dynamic and not at all static.

As a preliminary, I must make the obvious point, the European Community does not have an 'agricultural surplus' according to the internationally accepted definition of 'agriculture'. We are, by far, the world's biggest importer of agricultural and food products - for example we imported more than 50 thousand million ECUs worth in 1983. We are the world's second largest exporter of these goods - around 27 thousand million ECUs worth in 1983. This means that we have a substantial trade deficit of around 25 thousand million ECUs in agricultural and food products, a figure which is approximately the same size as the total trade deficit of the European Community. That is to say that if we were self-sufficient in agricultural and food products, the Community would not be running a balance of payments deficit with the rest of the world.

However, Europe enjoys a relatively high standard of living due to 'free trade' and cannot set itself a target of 'self-supply'. I want to avoid the possible confusion that 'surplus' = 'exports' or 'surplus' = 'production - domestic demand' or 'surplus' = 'production + imports - exports + unsubsidised domestic demand'. I am trying to highlight already that the title 'Surpluses ...' is a play on words - a field to be cultivated by politicians, a quagmire to entrap the unwary. The apparently harmonious picture of no 'agricultural surplus' changes sharply when we bring into focus the various commodity sectors.

Table 1   EEC Trade Balance in Agricultural and Food Products

(Million ECU)

	1978 Exports	1978 Imports	1978 Balance	1982 Exports	1982 Imports	1982 Balance
All products	221 319	227 275	-5 956	280 671	314 972	-34 301
Agricultural and food products	16 933	46 038	-29 105	25 057	46 630	-21 573
Food products & live animals	10 852	26 734	-15 882	17 314	27 806	-10 492
Live animals	212	505	-293	400	583	-183
Meat	1 003	2 116	-1 113	1 727	2 446	-719
Milk & eggs	2 264	542	1 722	3 868	655	3 213
Fish	533	1 853	-1 320	727	2 357	-1 630
Cereals	2 274	2 987	-713	4 027	2 456	1 571
Fruit & vegetables	1 131	6 879	-5 748	1 661	7 462	-5 801
Sugar & honey	1 166	1 096	70	1 852	1 023	829
Coffee, cocoa, tea, spices	1 060	7 510	-6 450	984	6 136	-5 152
Animal feed	625	3 084	-2 459	1 141	4 479	-3 338
Other food products	583	162	421	928	210	718
Beverages & tobacco	3 582	2 488	1 094	4 802	2 227	2 575
of which: alcoholic beverages	2 937	715	2 222	3 802	782	3 020
Hides	417	1 428	-1 011	552	1 423	-871
Oilseeds	28	3 790	-3 762	26	3 778	-3 752
Natural rubber	8	698	-690	6	634	-628
Timber & cork	258	5 181	-4 923	355	4 986	-4 631
Natural textile fibres	260	2 796	-2 536 .	295	2 767	-2 472
Agricultural raw materials	722	1 106	-384	906	1 225	-319
Oils & fats	766	1 810	-1 044	737	1 772	-1 034

The first thing which is apparent from Table 1 is that we are dealing with a dynamic situation. While the total trade deficit has deteriorated, that for agricultural and food products has narrowed substantially. The reduction in this trade deficit is not apparently due to a reduction in imports but rather to an increase in exports of agricultural and food products. When one focusses on the various groups of commodities it becomes apparent what is happening - with one major exception. Our imports reflect a stable demand for various commodities which are not normally produced or which it is uneconomic to produce in sufficient quantities of the required quality within the European Community. The major exception is imports of animal feed to which I will return later. On the other hand our exports have expanded significantly both for livestock, livestock products and for the arable crops which we have a vocation to produce.

This is the result of the steady expansion of our agricultural production which in recent decades has increased at between 1.5% and 2% each year. At the same time, domestic demand has either increased marginally by no more than 0.5% each year or latterly stagnated. The inevitable conclusion must be that for many commodities (for example, milk, beef, cereals) effective support for the domestic market is only sustainable against a backdrop of buoyant world demand. Now that solvent world market demand is either stagnant or declining, we have surplus production which we either have to dispose of mainly on our domestic market by one means or another or which we have to buy into intervention.

We can now focus on a number of commodities and observe in greater detail what has happened to the balance of domestic supply and demand for most of the commodities which we produce in significant quantities. Let me start by giving a list of a number of commodities and then I will provide an explanation of the list:

    Milk
    Beef
    Cereals
    Tomatoes
    Currants and sultanas
    Rapeseed
    Sunflower seed
    Sugar
    Cotton
    Wine
    Olive oil

Now this list represents those market regimes for which the Commission has proposed, and the Council has accepted, the principle of 'guarantee thresholds'. I should explain what a guarantee threshold is: it is an (annual) level of production up to which the Community is committed to full market support. When production exceeds the guarantee threshold, action is taken appropriate to that common market organisation. In passing I should note that this inventory of market organisation represents well over half of European agricultural output.

If we pose the question "According to what general principles did the Commission decide the levels and means of applying the guarantee thresholds for each of these sectors?" it is rather difficult to provide a simple answer - despite my own involvement in the process. As regards the levels fixed, no explicit relationship was established nor has since been established between for example 'guarantee thresholds' and Community levels of self-supply. Indeed it may be noted that when taking decisions the Council of Ministers has emphasised that this is not the case. Nevertheless, it is apparent that 'guarantee thresholds' have been set for market sectors where public expenditure was liable to increase rapidly and the cost-effectiveness of FEOGA expenditure could be called into question.

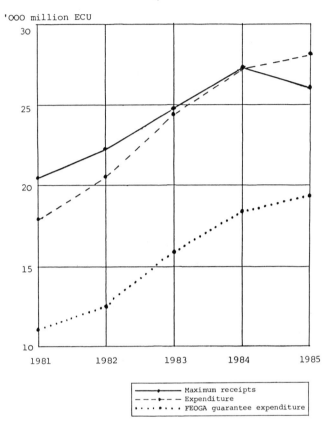

'OOO million ECU

Fig. 1. Budgetary Developments

Figure 1 illustrates with some force why the cost effectiveness of FEOGA expenditure has become such a sensitive issue. While all industrialised countries support their agriculture, increasing attention is being paid to the cost-effectiveness of measures designed to support, in the final analysis, producers' incomes.

One reason why FEOGA expenditure is increasingly coming under restraint is shown in Figure 2. Twenty five years ago farming gave direct employment to nearly 20 million people and only 2 million people were unemployed. Now farming employs far fewer people while the ranks of the unemployed have swollen to a level which gives rise to very serious concern. Action to contain or reverse the rising tide of unemployment costs money - but the total budget available for social, regional or integrated development programmes is limited, first by the restrictions imposed on public expenditure and second by the amounts allocated to agriculture.

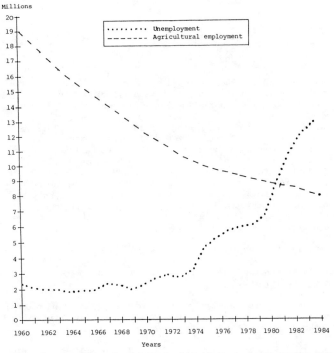

Fig. 2. Unemployment and Agricultural Employment in the EEC

As we turn to individual market organisations I have here a set of figures which illustrate the dilemma facing those who feel that there is an easy solution to the problem of surplus crop production - that we should grow more of what we import and less of what we export. The figures, which represent estimates of the cost to FEOGA per <u>hectare</u> of land cultivated in 1984, are as follows:

Common wheat	185 ECU
Barley	260 ECU
Rapeseed	302 ECU
Sunflower seed	340 ECU
Lupins	400 ECU
Peas and field beans	452 ECU
Lucern	491 ECU
Soya beans	682 ECU
Cotton	744 ECU

The corresponding average figures for the previous three harvests (i.e. 1981, 1982 and 1983) were:

Common wheat	307 ECU
Barley	223 ECU
Rapeseed	443 ECU

Sunflower seed	483 ECU
(Lupins - no support)	
Peas and field beans	262 ECU
Lucern	350 ECU
Soya beans	553 ECU
Cotton	792 ECU

In other words, if half a million hectares were switched from wheat to rapeseed, and half a million hectares were switched from wheat to sunflower seed, the million hectares reallocated would, on 1984/85 costings, add (302 - 185) x 1/2 + (340 - 185) x 1/2 million ECU = 136 million ECU; the figure based on the average of 1981, 1982 and 1983 is (443 - 307) x 1/2 + (483 - 307) x 1/2 million ECU = 156 million ECU.

In other words a simple switch from the cultivation of the more traditional cereals to the protein-rich crops such as rape or sunflower would, unless there is a change in policy, add to the cost of FEOGA and not significantly reduce the imbalance which exists in the cereals sector. It should be noted that 1 million hectares of cereals corresponds to between 3% and 4% of the area devoted to cereals, and since improvements in yield are averaging 3% per annum the effect of converting 1 million hectares each year to protein crops would:

- increase FEOGA expenditure;

- merely arrest the deterioration of the cereals market without actually improving it;

- disrupt the delicate balance of domestic protein production (the current areas allocated to rapeseed and sunflower seed being 1 million hectares and a 1/2 million hectares respectively).

Let us now examine the question of agricultural surpluses in a systematic way, starting with livestock and livestock products and then returning to crops. The following analysis which is not exhaustive, refers to the situation in the Community of '10'. While the progressive integration of Spain and Portugal into the Community will, of course, result in a different balance of supply and demand for various agricultural products, it will not eliminate existing surpluses.

I will start with the milk sector because here at least the situation is very clear - Europe has a surplus of dairy products, the historical background is well known. However, one historical point must be made, the transition from deficit to surplus was achieved with a constant dairy herd, there were 25 million dairy cows in the '10' in 1960 and there were still around 25 million dairy cows in the '10' at the end of 1983. Enormous gains in productivity produced steady increases in milk production. The market organisation which we had created, guaranteed the basic return to the producer and he was given little incentive to adapt to the market - instead he was incited to expand. Figure 3 shows how production was increasing while consumption stagnated.

At last, on 31 March 1984, the Council accepted the need to put a limit on dairy production and the era of milk quotas began. The decision was a hard one, but how else could one cope? Production was increasing at 3 to 4% per annum; domestic demand was stagnating (despite the various subsidised disposal schemes); exports were falling in a shrinking world market (with little prospect of recovery) and stocks were rocketing. By the end of March 1984, public stocks of butter were rapidly approaching the 1 million ton mark.

Million t.   EUR10

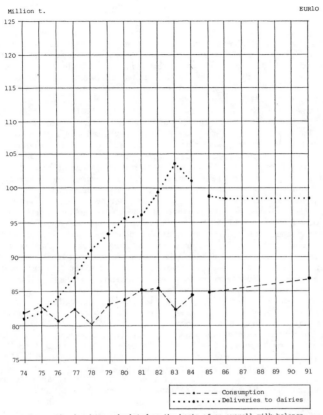

Consumption has been calculated on the basis of an overall milk balance
in milk equivalent terms (by reference to butterfat content)

Fig. 3. Milk

        The principle of the 'guarantee threshold' received a new and
sharpened interpretation. Producers exceeding a reference quantity would be
taxed by a super-levy at a rate designed to deter farmers from exceeding
their reference quantity. The system is working. Milk production in the
12 months from 31 March 1984 not only did not increase, it actually dropped
back to a less unreasonable figure. This was brought about by a combination
of two factors. The use of less-intensive feeding methods reduced farmers'
costs and reduced output; in some cases farmers' incomes rose as they
adapted to a more cost-conscious approach. Farmers culled more cows and for
the first time a significant reduction in the Community dairy herd has been
observed. In addition, various national schemes were implemented with the
common objective of persuading some farmers to stop milk production
altogether. However, the benefits of these national schemes - released
reference quantities - were promptly transferred to other dairy farmers in
an effort to alleviate the shock of transition.

However, we still have a dairy surplus and an increasing butter mountain, despite the efforts made to destock the surplus over the last 12 months. It is clear that in the foreseeable future the total amount of milk we produce will have to be reduced. The Commission will be reporting to the Council of Ministers later this year on the possible cost-effectiveness of a Community buying-out scheme. However, since the preparatory work on this has only recently begun it would be premature to comment on what conclusions the Commission might draw or what proposals the Commission might submit to the Council of Ministers.

When we turn to the beef sector we are faced with a little less gloomy picture. Beef has lost a significant share of the meat market while the overall domestic demand for beef has remained fairly constant. The Commission has the power, which it uses, to adjust within certain limits the market conditions for producers. The Commission has the flexibility necessary to absorb the variations in the beef production cycle, although this can be at considerable expense. Stocks of beef in intervention have risen considerably over the last year. Since most European beef derives from the dairy sector it is anticipated that the adjustments taking place in dairying will alleviate the pressure on the beef market. With flexibility, the outlook for the beef sector is reasonably healthy in the medium term. This is why the Commission has considered that implicitly its management powers - used responsibly - provide an effective 'guarantee threshold', without the need to put a precise figure on the threshold.

Let us now turn to sugar the classic crop for which potentially a surplus exists. Production exceeds consumption. We have 'A' quotas, 'B' quotas and C sugar and the plain fact is that sugar beet production is so attractive that no producer aims to risk producing under his quota. This means that in a normal year he will exceed his quota and in a 'good' year he will substantially exceed his quota. This is a clear conclusion of a little elementary micro-economic analysis - it is so gratifying when occasionally practice confirms theory. The surplus production is intended to be disposed of entirely at the cost of the producers. With a depressed world market - prices fell to 2.56 cents a pound in New York last week - the unpaid bill due by producers now exceeds 500 million ECU. It is clear that here again the Community has a surplus, a manageable surplus, but nevertheless a surplus. Having said that, I leave you to guess at what proposals the Commission will make to the Council for the next five years of the sugar regime.

I will quickly gloss over tomatoes, currants and sultanas, and cotton by saying that the substantial cost of each of these regimes has persuaded the Council to implement a system of 'guarantee thresholds'. In each case, Community support for the products in question is affected when production exceeds a predetermined limit. This is the response of the Council to the explosion in expenditure on so-called 'Mediterranean products'. It leaves intact the problem of the tobacco regime, but apart from 'olive oil', where support is restricted to 'traditional' producers, and wine, which merits special attention, action has been initiated for all the so-called Mediterranean products.

For wine a measure of the scale of the problem is that production is about 150 million hectolitres and consumption is about 130 million hectolitres. After much soul-searching, the Council accepted a two-pronged approach to the problem of the growing structural surplus which exists in the table wine market. First a significant step forward was made by the introduction of a grubbing up and abandonment programme which should restrain or curtail the supply side - this will, of course, cost money over a number of years. The second and parallel step forward was the dilution of price support, with a very significant reduction in the price paid for wine which is compulsorily distilled - this should restrain the growth in expenditure in the short and medium term. On balance, it is hoped that the return to a more balanced market for table wines will ensure that producers' incomes are largely supported by the market and that expenditure in the sector remains within reasonable limits.

This change of policy which included allusions to a 'guarantee threshold' of 100 million hectolitres of market prices, to the level of stocks, to market prices, widened the concept of guarantee thresholds to include considerations of a financial nature. A balance between expenditure now and future expenditure had been struck; gone was the static approach - can we afford it today, for tomorrow we can always review the situation. Incidentally I should mention in passing that the stocks in the table wine sector are almost exclusively carried at the cost of the producer, and so here we have had a surplus without the appearance of a wine-lake to rival the butter mountain. Nevertheless, the market instrument of distillation which removes surplus table wine from the market is very costly.

Now I will turn to the crops: <u>cereals</u> and the domestically produced substitutes <u>rape</u> and <u>sunflower</u> seeds.

Since 1980, the Community has systematically produced more <u>grain</u> than it consumes. On the one hand, domestic demand for grains is virtually stagnant, there is little prospect of any significant increase in domestic demand for human consumption while demand from the livestock sector is at present subject to two factors.

First, the livestock sector is in stagnation, the impact of milk quotas has reduced demand for compound feed for dairy cattle for the first time in more than twenty years. Demand by the rest of the livestock sector follows the production cycles of the beef, pigmeat and poultrymeat sectors which gives an outlook of very limited increases in demand for animal feed during the rest of this decade. Indeed the short term picture is even more bleak. Estimates for 1984 imply that compound feed production fell by 2% to 3% compared with the previous year - a reversal of the trend of regular increases of 3% to 4% observed over the previous decade.

The second factor is the competition from so-called 'cereals substitutes' whether produced within the Community or imported. While Community cereals prices remain at their present level the most optimistic outlook is that the uptake of cereals for feed will stagnate.

There is also the risk of surpluses developing for <u>rape</u> and <u>sunflower</u> seed.

Understandably the Council of Ministers has accepted the need for 'guarantee thresholds' for <u>cereals</u>, <u>rape</u> and <u>sunflower</u> seeds. The necessity for parallel action on <u>rape</u> and <u>sunflower</u> seed, even before a surplus has occurred, stems from two reasons to which I have already alluded. First, the marginal unit cost of support for these regimes is far higher than for cereals. This extremely high cost, in large part, reflects the absence of Community preference on imported 'cereals substitutes'. The second reason is that if there were any significant conversion from cereals into these crops, not only would budgetary expenditure rise but the markets for colza and rapeseed would soon become swamped.

The mechanism which they chose for all of these products was price. Production in excess of a predefined limit would be translated into a reduction in market support for the following season.

Why the price mechanism? Simply because there is little alternative. The services of the Commission have looked at the possibilities of 'set-aside' and of 'quotas' and do not find that they are very attractive in a <u>European</u> <u>environment</u>.

The services of the Commission are looking at the possible new outlets for what is currently surplus production. It is estimated that there are possible new outlets, but insufficient to balance the books.

With cereals yields rising at around 3% per annum, unless something is done and quickly, the harvest of 1984 - over 150 million tons of cereals - could become the norm by the end of the 1980s. It took only one harvest of that scale to push up intervention stocks by around 10 million tons compared with the previous year. That they rose by so little was due to a combination of factors which are unlikely to recur with any regularity.

The inescapable conclusion is that cereals production must be rendered less attractive, and that parallel action must be taken on rape and sunflower seed. The Commission will soon be publishing a discussion paper on the perspectives for European agriculture. What to do about the cereals surplus and how various options would affect farmers will be an important feature of the forthcoming debate.

It is not my purpose today to discuss what conclusions the Commission should draw from the debate on the perspectives for the CAP. Whatever adaptation of existing policy the College may conclude is necessary, I am convinced that we will have to deal with surplus production for the rest of this decade. Change rarely occurs overnight and the issues at stake militate in favour of gradual adaptation - evolution rather than revolution.

The work of this seminar can provide a useful contribution to the thinking of the Commission and the public debate which will follow publication of the Commission's 'Green paper'. However, I am convinced that the work which you are doing will also have a longer term significance.

FUEL PRODUCTS OR ADDITIVES ORIGINATING FROM AGRICULTURE
COMPARISON OF THE VARIOUS TECHNICAL SOLUTIONS

H. QUADFLIEG

Technischer Überwachungs-Verein Rheinland e.V.
Cologne, FRG

1. INTRODUCTION

For fuels and fuel additives crude oil is worldwide the predominant source for the time being. The events of 1973 and 1978 indicating possible supply shortage and price explosions for gasoline and diesel fuel have stimulated interest and research in the field of alternative fuels. Those based on renewable energy sources seem to offer particular attractiveness, since solar energy is free of cost and the biomass availability is unlimited in comparison with fossil resources such as crude oil, coal or natural gas.

However, except in Brazil (Proalcool) and USA (Gasohol), a market introduction of biomass fuels has not yet occurred to any great extent. The high cost of agricultural fuel products originating principally from the low energy density of biomass has not led to economic competitiveness with conventional or other alternative fuels. Figure 1, with regard to market prices for different fuels or fuel components, shows that biomass products rate at least double the price of gasoline.

Progress in the overall economic situation can be expected primarily from research and development in the fields of plant growing and biomass product processing. Compared with the fuel production costs in excess of conventional fuels, the additional operational costs caused by modified relevant technical concepts in the transport sector are relatively low. In the initial phase, however, high investments, depending on the type of biomass fuel application, may be needed there, too.

Comparing the various technically possible solutions[1] emphasis will be put particularly on the fact that Western Europe suffers from a surplus of agricultural products for which, at present, no sufficient outlet into the market can be found.

2. FUELS FOR ROAD TRANSPORT

Worldwide present and possible future use and demand of conventional and new fuels are illustrated in Figure 2[2] and Figure 3[3]. With respect to regional availability biomass fuels, also in the future, will have to compete with other favourable energy sources. Clearly, the best chance for production of energy from biomass has to be attributed to tropical and subtropical areas (cf. Figure 3).

3. BIOMASS FUELS

Favourable production possibilities for fuels from renewable raw materials are shown in Figure 4[4]. Other processing lines such as the hydrolysis of cellulose materials e.g. to methanol, or the use of gaseous fuels from biomass fermentation, or wood gasification, will not be considered here, since their competition levels are still relatively low. However, regional application of biogas e.g. in farm tractors, should generally not be excluded in such cases where biogas will be produced for other reasons anyway. One of the main problems of gaseous fuels arises from the high storage capacity (volume) needed to gain a reasonable mileage for vehicles.

Methanol, as a biomass fuel, will not be regarded here in more detail, since it can be produced much more economically from fossil resources. Incidentally, experience and know-how about methanol application in road vehicles is available worldwide to a higher extent than for any other alternative fuel[1,5].

Thus, ethanol and vegetable oils or its derivatives, respectively, are the most interesting biomass fuels. Today, ethanol production relies almost entirely on sugar or starch containing materials. However, for long term and high level substitution aspects, ethanol should be based more on cellulose feedstocks, as they denote a much higher portion of the totally available biomass and as they are not a link in the production line for human food.

4. VEGETABLE OILS

From Figure 5[3] which shows the suitability of various alternative fuels for Otto and diesel engines, it can be seen that vegetable oils are not appropriate for use in spark ignition engines. This is due to the relevant boiling behaviour shown in Figure 6[3] for different fuels.

From the fuel data comparison in Figure 7[3] the high viscosity and the poor behaviour at low temperatures of vegetable oils are recognised, whereas ignition quality and energy content are comparable to the diesel fuel data. Clogging of injector nozzles (Figure 8[6]), contamination of fuel filters, oil thickening and extensive odour of the exhaust gas are further handicaps which can largely depend on the type and saturation degree of the various feedstocks (cf. Figure 5).

These problems can be reduced by blending with diesel fuel or by transesterification of the vegetable oils. The resulting monoesters (methyl and ethyl ester) show a high degree of compatibility with existing motor technology[7]. Figure 8[6] shows that the viscosity will be decreased significantly. Also cetane numbers are improved (compare Figure 7). But there are problems left to solve for the application of vegetable oil esters (as unit fuels) in unmodified or only slightly modified vehicles: insufficient resistance of some materials, change of fuel composition during storage, dilution and, later, thickening of lubricants with effects on engine durability[3,7].

The substitution of diesel fuel by vegetable oils or their derivatives is considered likely to become much more significant for developing countries than for Central Europe due to the fact that, worldwide, a shortage of middle distillates is prognosticated and that there is a lack of relevant infrastructures for goods traffic in many areas. In a first phase probably road transport by diesel engines operating on domestic fuels has to contribute to solve those problems.

5. ETHANOL FUEL AND ENGINE CONCEPTS

Ethanol, from its characteristics, particularly by its high octane and low cetane number (below 10) is, as a unit fuel, an ideal substitute for gasoline. In Figure 9[9] some fuel data of gasoline, methanol and ethanol are compared. But ethanol can also be applied in diesel engines (cf. Figure 5). In addition to the potential to substitute conventional fuels, ethanol can also bring advantages to factors such as fuel consumption efficiency or environmental protection.

In the following the amount of ethanol fuel content and the resulting extent of necessary engine modifications are selected for concept classification, parameters which are directly linked to the overall possible substitution.

5.1 Low Blends without Engine Modifications

The primary advantage of low ethanol admixture to conventional fuels is that the blending component does not imply additional costs on vehicles and fuel service structures and that from a technical view a market input would be possible any time.

Such concepts have been realised to a larger extent only for gasoline, since for diesel fuel the necessary efforts to avoid negative effects (phase separation, cetane number decrease) are higher than the advantage of a small diesel oil substitution. Indeed, initially it was the substitution ideas which led to E20 in Brazil and to Gasohol (10% ethanol) in the USA, and additional aspects like octane enhancing, particularly for unleaded gasoline, came into consideration later.

There are technical and other limits for the use of ethanol and various oxygenates as gasoline extenders. The proposal for a relevant EEC directive is still under discussion. For ethanol the maximum blend limit will probably be 5% and for the overall oxygen content 3.7% is envisaged on the condition that the prescribed limitations for the single components are not violated (e.g. 3% for methanol). The dependence of gasoline oxygen content on blending component amounts is shown for different oxygenates in Figure 10 [10].

With regard to European significance the low percentage admixture concepts will be discussed in more detail in section 6.

5.2 Blended Fuels for Modified Engines

Anhydrous ethanol can be blended with gasoline without difficulties. In the presence of water, cosolvents or high aromatic content of the gasoline are suitable measures against phase separation. Since ethanol and other oxygenates, particularly at higher blending rates for achieving a reasonable lengthening of conventional fuel supply, have remarkable effects on quality criteria of the basic fuel, modified or adapted engines are necessary as well as separate service lines including pumps.

Various investigations on alcohol-gasoline blend fuels - more for methanol because of economic reasons - have demonstrated the general technical feasibility of appropriately optimised concepts [1,5,11]. For blend fuels, e.g. E50 or E20 (in Europe not acceptable for unmodified present cars), early agreements on a European level and high investments would be needed before an implementation could start. Although such fuels could improve some fuel characteristics, e.g. octane numbers or exhaust gas emissions, they do not make optimum use of ethanol's properties. Hence, this kind of blend fuels may receive concrete interest, if at all, particularly during a transient introductory phase or in emergency cases. This is also true for flexible engine concepts which use sensors to determine the ethanol content during operation and thereby adjust the correct air fuel ratio; optimisation for the total range from gasoline (E0) to ethanol (E100) is not possible.

The solubility of ethanol in diesel oil is poorer than in gasoline and deterioration of blend stability is observed here also with decreasing temperature and increasing water content. Usually higher efforts and expenses are needed for the engine and/or fuel technology adaption. Since the emission behaviour is more complicated for diesel engines it may be that emission control aspects will lead to more interest in alcohol-diesel blend fuels.

In the following figures [12] e.g. some results from investigations on three different blend fuel concepts are selected. In Figure 11 the composition of the blends is shown, in Figure 12 the cetane number decrease for two different solubilisers with increasing ethanol content. Figures 13-16 show improvements for the most crucial exhaust gas components and reasonable fuel economy in comparison with plain diesel fuel.

There are other concepts for fuel mixtures which have been investigated. By the mechanical production of a diesel-ethanol emulsion just before entering the combustion chamber fuel stability has to be realised only for a short period, the mixing device may be complicated, however.

It is also possible to add ethanol to diesel fuel up to 30-40% by aspiration (fumigation) through the air intake with a carburettor or low pressure injection. A load and speed dependent admixture control can be favourable with this technique.

5.3 Ethanol Fuel (E100)

There is no concept which really uses 100% ethanol. Hence, those techniques will be denoted as E100 which operate essentially (minimum ethanol contribution 85-90%) on alcohol. The lower energy content (calorific value) and the higher evaporation heat of alcohols compared with conventional fuels (Figure 17[1]) are the basic reasons for necessary changes and optimisation of present engine technology. This can work generally in three ways:

- on the fuel side, e.g. cold start additives for Otto engines or ignition improvers for diesel engines,
- by dual fuel systems, e.g. using a pilot ignition fuel in diesel engines,
- on the engine side by special devices, e.g. electric heating for cold start or spark plugs in diesel engines for ignition assistance.

5.3.1 Otto Engines

The high octane numbers (Figure 18[1]) and evaporation heat (cf. Figure 17) of alcohols enable an efficiency improvement for spark ignition engines by increasing the compression ratio. Figures 19[9] and 20[13] show the good performance and the positive high compression effects of ethanol and methanol. The lower calorific value needs adjustment of fuel flow rates and air/fuel ratio in conventional engines.

Much more care has to be taken for the mixture formation in alcohol operated engines due to the higher evaporation heat, the different volatility and the lower vapour pressures of alcohols. Figure 21[1] shows that the low vapour pressure causes cold start problems, particularly for ethanol.

The cold start behaviour can be improved by additional heating devices for fuel and air, by low boiling hydrocarbon additives (isopentane, gasoline, butane), by increasing the ignition energy, by thermal or catalytic splitting of ethanol into gaseous components. Pre-ignition problems can be solved for ethanol more easily than for methanol.

Alcohols are clean burning fuels and generally show lower pollution effects on air, soil and water. Due to the evaporation heat - especially NOx - emissions are relatively low. Figure 22[1] shows, schematically, the exhaust gas emissions of an alcohol engine related to a corresponding gasoline series product motor. Only the aldehydes are higher with alcohol operation. These can be reduced significantly by an exhaust gas catalyst as Figure 23[13] shows.

Practical experience with ethanol engines is available. Also, the same fundamental motor concepts are suitable for both methanol and ethanol with relevant adjustments. Alcohol resistant materials or material protection (e.g. coating) are known and available without significant extra costs.

Special lubricants have been developed to reduce problems with deposits in alcohol engines. Single corrosion problems are mainly due to insufficient fuel qualities (acid content or contamination of ethanol too high).

Figure 24[1] tries to sum up the main quality variations of alcohol fuels from the corresponding gasoline features.

5.3.2 Diesel Engines

The low cetane number of alcohols (cf. Figure 18) indicates that they are not an ideal fuel for compression ignition engines. Neither is there such a possibility to increase efficiency as there is with Otto engines, since the diesel concept is already so effective and the higher evaporation heat of alcohols does not support self-ignition. Under good conditions some alcohol concepts may reach a level of energy efficiency about equal to diesel fuel operation (Figure 25[13]).

It is the above mentioned substitution demand, particularly in developing countries, and the soot-free burning of ethanol and methanol which, nevertheless, has stimulated the development of various concepts for alcohol operation of diesel engines.

Ethanol with ignition improvers

Ignition quality is achieved by ignition improvers which are usually organic nitrates. Among those, TEGN (triethylenglycoldinitrate) which can be produced from ethanol is especially well suited. Necessary amounts, costs, availability and environmental compatibility will finally influence the selection of appropriate improvers. Figures 26 and 27[14] show the relevant relationships for various products. The engine itself needs only relatively small modifications, a readjustment for diesel oil operation seems possible without much effort. Such an ethanol fuel is, however, not applicable to Otto engines. About 1000 vehicles using this technique are under operation in Brazil.

Dual-fuel injection

Diesel oil is used in small amounts (10-15%) as a pilot fuel to secure ignition for each engine cycle, the main fuel being alcohol. The vehicle needs a double fuel system with two storage tanks and two injection pumps (or a specially controlled one) which causes additional expenditure. This concept relies widely on known conventional technology and, because of the complicated handling with a double fuel system, will probably not be the final optimum choice.

Assisted ignition

To overcome the poor ignition quality of alcohol, the engine will be equipped with ignition components such as spark - or glow - plugs or hot spots which require change of the engine head and additional costs. It will not be discussed here whether one comes up with a diesel or Otto engine. In any case, the concept is based on an original diesel motor. The sensitive area may be lifetime and/or energy demand of the plugs, items which are linked to maintenance and service intervals. One advantage is that this type of engine can use fuels which are also compatible with Otto engines.

5.3.3 Dissociated Alcohols

To use the waste motor heat for splitting alcohols into gaseous components (methane, CO, H_2) and operate the engine as a gas-motor (spark-ignition) seems attractive with respect to good fuel consumption and clean exhaust gas. However, a lot of additional equipment is needed for vehicles with either Otto or diesel engines. A methanol version based on a diesel motor has been developed to such a level; that operation is ongoing in several buses under normal traffic conditions. This principle, with a liquid fuel for favourable storage and a gaseous one for clean combustion, needs more basic research in order to learn how far the theoretically possible advantages can be transferred to practical application, particularly for Otto engines.

5.3.4 Comparison of Diesel Concepts

Alcohols, being principally not an ideal diesel engine fuel, require more or less additional efforts either on the fuel or the engine side. All concepts show selective advantages, and an assessment of distinct priorities is not yet possible. Only the dual-fuel system seems to have minor chances under operational aspects. For some of the concepts more experience is available with methanol than with ethanol.

All these concepts show significant advantages over the diesel oil operation with respect to the cleaner exhaust gas. Figure 28[4] shows one example. The combustion of alcohols reduces particulate emissions almost completely. NOx emissions are lower (high air fuel ratio of diesel engine) and PAH (polycyclic aromatic hydrocarbons) are non-existent. Further development of such concepts may be influenced by the future demands on emission levels for commercial vehicle engines.

6. INTRODUCTION OF ETHANOL IN EUROPE

6.1 Substitution Potential

Independent of ethanol production economy, the low admixture concept for unmodified cars is the most favourable way for ethanol to enter the fuel market in Europe. Following the arguments discussed in section 5.1 and referring to the 5 Vol.% limit of blending envisaged by the EEC directive (proposal), the maximum substitution results from the gasoline demand in Western Europe. Figure 29[2] gives a survey on present and future crude oil demand for 14 countries. The possible ethanol consumption amounts to:

Year	Motor Gasoline Demand Mill. t	Ethanol Potential (5 Vol.%) Mill. t
1979	82.9	4.35
1985	77.1	4.08
2000	67.1	3.55

Without sticking too closely to the prognosticated trend of gasoline demand, one recognises that 4 million tons of ethanol can be regarded as a reasonable order of magnitude to start with.

If a European-wide introduction of ethanol as a blend or unit fuel for modified cars is to be discussed, assuming this would be politically and/or economically meaningful, then for technical and organisational reasons the E100 route should be preferred. In a similar way, what has been investigated already for methanol would also be true for ethanol[5,11]): a combination of E100 and E5, where the flexibility of E5 helps to buffer between E100 production and E100 demand (supply guarantee), would probably be most favourable.

6.2 Technical Aspects of Oxygenates for Gasoline

Oxygenates (cf. Figure 17) have been known for a long time as generally suitable gasoline components. Figure 30 shows relevant fuel parameters which can be positively or negatively affected by oxygenate blending and thereby influence the engine behaviour. Primarily, ethanol will be discussed in the following. The possible indirect input of ethanol in the form of ethyl tertiary butyl ether (ETBE) will only be mentioned. ETBE has a higher octane enhancing quality than ethanol but not quite as good as MTBE (based on methanol) which is already used in the fuel market. The production of both ethers requires isobutane, the availability of which is limited at present.

Octane numbers

The stepwise phasing down of lead content and the intended introduction of unleaded gasoline in Europe recently have increased the interest in oxygenates again. The enhancing effect of lead on the research octane number (RON) and the motor octane number (MON) is shown in Figure 31[2] for premium and regular gasoline. From Figure 32[15,16] where the influence of methanol and ethanol on the octane numbers is plotted, it is clear that neither alcohol is able to compensate the lead, but they have a positive effect which is a little bit better for methanol than for ethanol. The increase of RON (acceleration) is higher than that of MON (high constant speed). Comparing premium and regular, we see that the increase is higher for a lower starting octane level (regular). This is also confirmed by the curves shown for the oxygenate effects on an unleaded gasoline.

These octane effects, even if partial in comparison with lead, unfortunately are not reflected by the behaviour of real engines at full load and full speed. The street octane number (SON), corresponding to this condition, can decrease with ethanol and methanol admixture as Figure 33[16] shows. This trend was only observed at high rpm and depends strongly on engine types. None of the engines tested confirmed, however, the positive results gained for the laboratory octane numbers (Figure 32). At best, some engines do not react in a negative sense.

It has been shown, e.g. for methanol, that the SON behaviour can be corrected, at least partially, by adjustments of the air fuel ratio. Figure 34[10] shows a diverse 'octane requirement shifting' by 7% methanol admixture dependent on the basic rich or lean mixture formation.

Volatility

How deviations from the 'normal' boiling curve will affect engine performance (adjusted to 'normal' fuel condition) is demonstrated in Figure 35[2]. If we now look at the volatility changes caused by 10% admixture of different oxygenates in Figure 36[15], we note a remarkable volatility increase by methanol and ethanol in the temperature range between 50° and 100°C. Hot driving troubles and vapour lock may result in unmodified engines, which becomes evident from the upper part of Figure 37[16]. In a test with a small vehicle fleet (seven cars) on various oxygenated gasoline fuels we see that hot driving problems indeed dominated.

For various gasoline components, a conflict arises between octane enhancing and volatility changing effects. Figure 38[2] shows this for the example of reformate mixed with a light catalytically cracked product. Because of such effects there are limits also for alcohol blending with gasoline (volatility too high) just independent of other limitation criteria (e.g. material resistance).

Vapour pressure

Ethanol and methanol have a lower vapour pressure than gasoline (cf. Figure 21). As gasoline components, however, at low admixture rates they increase the vapour pressure of the blend fuel, which is shown in Figure 39 for methanol, ethanol and TBA. The volatility increase can be compensated by removing butane or adding components like IBA and/or TBA. Such combination restraint may reduce the maximum blend feasibility of single components, e.g. of ethanol.

Water tolerance

Fuels have to stay under all conditions in one homogeneous liquid phase. Oxygenates are very sensitive to water. The water absorption capacity increases with temperature and blending amount. Good properties as cosolvents or stabilisers have ethanol, IPA, IBA and TBA. For a 10% admixture of these components we see the temperature of beginning phase separation in Figure 40[15]. For IBA and TBA it is not really the point of phase separation but of first cloudiness caused by sprayed water droplets.

In methanol containing gasoline these cosolvents have not all the same good effects. In Figure 41[15] the combined blending of methanol and TBA proves to be more effective on water absorption (synergistic impact) than that of methanol and ethanol (just adding effect). Thus, in the presence of methanol the cosolvent quality, particularly of ethanol and TBA, is levelled and since methanol probably cannot be removed for economic reasons, price and availability of the different components will finally decide the product selection[17].

Fuel consumption

Increasing oxygen content and resulting lower calorific value of oxygenates (cf. Figure 18) reduce the blend fuel energy content proportionally to the blending percentage. For low admixture rates a corresponding increase of fuel consumption usually is not observed. The lower graph of Figure 37[16] shows this fact for various blends with ethanol and methanol up to 10 Vol.%. It is caused by the different combustion properties of oxygenates and by its leaning effects on the blend air-fuel mixture.

It has to be observed that surplus fuel consumption due to oxygenates in gasoline, supposedly without any labelling, is not primarily a technical problem but can probably be one of consumer protection (same price for different energy contents).

Emissions

The middle part of Figure 37 shows typical variations of exhaust gas emissions due to the above mentioned effects of oxygenates on combustion. Depending on mixture formation devices and basic adjustments of the investigated vehicles, wide scattering of the emission data is observed. CO is reduced for all cars. On average, emissions of HC are equal and those of NOx are a little higher. It has to be expected that aldehyde emissions, which are low and not jet regulated for gasoline, will be increased in the presence of oxygenates.

Material resistance

Some oxygenates, particularly methanol, are aggressive to certain materials, to metals as well as to elastomers. That is why the limit of 3% methanol admixture is set in the EEC proposal. Ethanol is far less critical, but also not compatible with all materials used today in car manufacturing. The problem is not to find appropriate resistant materials, however, but to secure compatibility with present cars. For ethanol 7% admixture could be accepted, less if other oxygenates, particularly methanol, are simultaneously blended.

6.3 Fleet Test with 'Eurosuper'

At present, the German Federal Ministry of Food, Agriculture and Forestry is carrying out a fleet test with an oxygenated fuel denoted 'Eurosuper'. Relevant data of this investigation are shown in Figure 42. This project tries to find out how unmodified series production cars behave when operated with a premium gasoline fuel oxygenated up to the blending limits which are envisaged by the relevant EEC directive. The composition of 'Eurosuper' is shown in Figure 43.

As a preliminary result it can be summarised that no significant problems have arisen in usual vehicle operation up to now. The chassis dynamometer measurements confirmed the trend of the above mentioned results with respect to exhaust gas emissions and fuel consumption[18]. Compared with a reference fuel (3% methanol, 2% TBA) the effects of 5% ethanol could be eliminated to lower CO for all seven cars, lower HC for five cars, a little bit higher NOx on average (two cars equal, two cars better, 2 cars worse). Aldehyde emissions are higher, up to a maximum of about 40% (two cars). Fuel consumption is equal on average. Some vehicles are slightly better, some slightly worse.

7. CONCLUSION

Under short and middle term aspects the best potential biomass fuels are vegetable oils and ethanol. Various technical concepts have been investigated. The state of knowledge in fuel application technologies is so far developed that from the view of fuel and vehicle engineers no major objections exist against introduction into the road transport field.

For Europe, a low admixture of components to gasoline is the concept to choose at first for ethanol, which probably cannot be treated as an isolated application, but has to be assessed within the spectrum of other favourable oxygenates. The limits of ethanol blending with gasoline for operation in unmodified vehicles will depend - independent of economic conditions - on technical criteria.

Ethanol is no lead substitute for octane enhancing. In spite of the poor effect on the street octane number of existing cars, it is believed that, for the future, ethanol, like other oxygenates, will gain more credit, particularly in view of the restraints for the octane pool situation of unleaded gasoline.

As a cosolvent, anhydrous ethanol has some good basic properties. Again, however, it has to compete with other components showing even better overall fuel stabilising qualities (TBA). There is no general and final answer to the question, what price could be attributed to ethanol as a fuel or fuel component. Let me quote N. Rask[19] who, in his investigations, came to the following conclusion:

"In valuing ethanol as a potential domestic and export commodity, the three levels of use identified above are considered; anhydrous as an octane booster, anhydrous as an equal substitute for gasoline, and hydrous ethanol as a pure fuel. Each of these uses is priced in relation to gasoline as follows:

Type of Ethanol	Ethanol price relative to gasoline price
Anhydrous ethanol	
octane enhancer	1.2
gasoline substitute	1.0
Hydrous ethanol	
pure fuel	.8"

Although this result may reflect more in the US situation than under European conditions, one can regard it as a starting point for further discussions.

8. REFERENCES

(1)   MENRAD, H., KÖNIG, A., Alkoholkraftstoffe, Springer, Wien-New York, 1982.

(2)   REGLITZKY, A.A., DABELSTEIN, W.E.A. (1984). Trends in Future Automotive Fuels on the German Market, Europ. G.F.C.-Symp. Autom. Fuels and Lubricants, Paris, Oct. 3-4.

(3)   REGLITZKY, A.A., HALTER, H.J., KNAAK, M. (1985). Kraftstoffeinfluss auf die Motorenölprüfung, mögliche Entwicklungstendenzen, Tribologie und Schmierstofftechnik 1, p.16.

(4)   BANDEL, W. (1981). A Review of the Possibilities of Using Alternative Fuels in Commercial Vehicle Engines, Int. Conf. Energy Use Management, Berlin, Oct. 26-30.

(5)   QUADFLIEG, H., BRANDBERG, A., TIEDEMA, P. (1983). Alcohols as Alternative Fuels for Road Vehicles, Rep. Subgroup 'Alcohol Fuels' to Management Comm. COST 304, EEC, Brussels.

(6)   BRUWER, J.J., v. BOSHOFF, B.D., HUGO, F.J.C., FULS, J., HAWKINS, C., v.d. WALT, A.N., ENGELBRECHT, A. (1980). The Use of Sunflower Seed Oil in Diesel Engined Tractors, Proc. IV. Int. Symp. Alcoh. Fuels Technol., Guaruja, Brazil, Oct. 5-8, Vol. 1, p.397-402.

(7)   PISCHINGER, G., CLYMANS, F., SIEKMANN, R. (1981). Diesel Oil Substitution by Vegetable Oils - Fuel Requirements and Vehicle Experiments, Proc. Int. Conf. New Energy Conservation Technologies, Berlin, April 6-10, Vol. 2, p.1377-1389.

(8)   BACON, D.M., MONCRIEFF, J.D., WALKER, K.L. Alternative Fuel Options in the Diesel Engine, Proc. Vol. 2, p.1369-1376.

(9)   ABTHOFF, J., HÜTTEBRÄUCKER, D. (1982). Adaption and Optimization of 4-Stroke-Engines for Operation with Alcohol Fuels, Proc. V. Int. Alc. Fuel Technol. Symp., Auckland, May 13-18, Vol. II, p.1-8.

(10) DABELSTEIN, W., REDERS, K. (1984). Unverbleite Ottokraftstoffe - Möglichkeiten und Grenzen der Herstellung, Paper presented at Technische Arbeitstagung Hobenheim, Stuttgart-Hohenheim, April 4-6.

(11) BANDEL, J., HASELHORST, M., KROEG, F., MAJUNKE, H.J., MENRAD, H., NIERHAUVE, B., QUADFLIEG, H., SEIDEL, G. (1984). Alternative Energy Sources for Road Transport-Methanol/The German Fed. Min. for Research and Technol., Prod. TÜV Rheinland, Köln.

(12) WEIDMANN, K., MENRAD, H. (1984). Fleet Test, Performance and Emissions of Diesel Engines Using Different Alcohol-Diesel Fuel Blends, SAE Techn. Paper Ser. 841331.

(13) FÖRSTER, H-J. (1982). Survey of Engine and Vehicle Development in Relation to Alcohol Fuels, Proc. V. Int. Alc. Fuel Technol. Symp., Auckland, May 13-18, Vol. IV., p.67-99.

(14) HARDENBERG, H.O., SCHAEFER, A.J. The Use of Ethanol as a Fuel for Compression Ignition Engines, SAE Reprint 811211 from Sp-503-Altern. Fuels for Diesel Engines.

(15) GIERE, H.H., NIERHAUVE, B., GONDERMANN, H. (1980). Anwendungstechnische Untersuchungen von Ottokraftstoffen mit sauerstoffhaltigen Komponenten, Mineralöltechnik, 7 July.

(16) NIERHAUVE, B. (1985). Anwendungstechnische Erkenntnisse zur Einsatzmöglichkeit von Methanol und Ethanol als Ottokraftstoff-Komponente, Entwicklungslinien in Kraftfahrzeugtechnik und Strassenverkehr - Forschungsbilanz 1984/Bundesm. f. Forsch. und Technol., Prod. TÜV Rheinland, Köln, p.V101-V112.

(17) SCHLIEPHAKE, D. (1983). Ist Bio-Alkohol eine Lösung für bleifreies Benzin, Paper presented at Verbindungsstelle Landwirtschaft - Industrie e.V., Essen, Sept. 21.

(18) Verbrauchs- und Abgasverhalten von Pkw mit Ottomotoren bei Verwendung von Eurosuper, Test Rep. F 133 - 04/85, Lehrstuhl für Angewandte Thermodynamik, RWTH Aachen, April 1985. (Not yet published).

(19) RASK, N. (1984). Alcohol Imports to the U.S. Fuel Market, Proc. VI Int. Symp. Alcoh. Fuels Technol., Ottawa, May 21-25, Vol. II, p.353-359.

Market Price for a Rough Comparison (1984)		
Component	Price without Tax	
	DM/l	DM/GJ
Premium Gasoline	0.67	20.80
Diesel Fuel	0.62	17.60
Methanol	0.33	21.10
Ethanol	1.28	60.60
TBA	0.73	27.20
IBA	1.27	48.60
MTBE	0.73	28.00
IPA	1.47	62.20
Vegetable Oil	1.38*	39.40

* Refined Product on World Market

Fig. 1. Market Prices for Fuels and Fuel Components

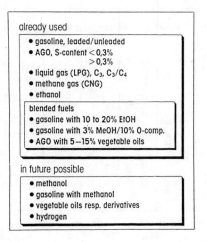

Fig. 2. Fuels for Road Transport Worldwide

Feedstock	Geographical Availability	Produced Fuels								
		P Gas	D F	Marine D F	Av. T F	C N G	L P G	Meth.	Eth.	Veg. Oil
Crude Oil (Domestic)	USA , OPEC Parts of S.- America Parts of Europe	•	•	•	•		•			
Crude Oil (Imported)	All , particular in Indust. Countries	•	•	•	•		•			
Tarsand , Shale	Canada , USA Australia	•	•	•	•					
Coal (Domestic)	USA , Europe Australia , S.- Africa	•	•					•		
Coal (Imported)	Europe Japan ?	•	•					•		
Biomass	S.- America , Africa Australia , Phillip. etc.							•	•	•
N G (Domestic)	OPEC , S.- America N.-Zealand, Thail.,USA Canada , Australia	• via Meth.				•	•	•		
N G (Imported as Methanol)	Europe , USA Japan							•		

Fig. 3. Fule Application Expected until the Year 2000

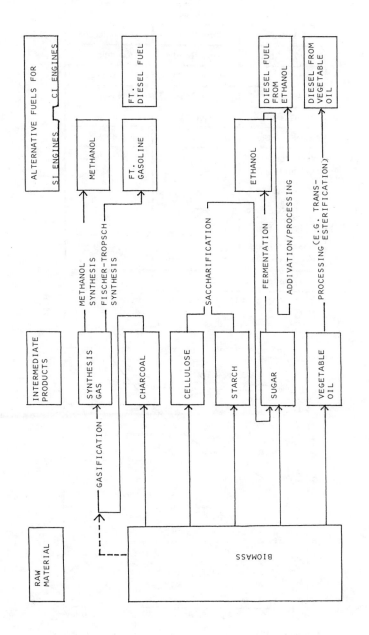

Fig. 4. Production Possibilities for Alternative Fuels from Biomass

Fuel	Otto Engine	Diesel Engine
LPG	Very Good	Conditional
LNG / CNG	Very Good	Conditional
Methanol	Very Good	Possible
Ethanol	Very Good	Possible
Alc-Gas-Blend	Good	Conditional
Alc-Diesel-Blend	No	Suitable
Vegetable Oils Sunflower Soya Peanut Rape Cotton-Seed Palm Coco-Nut	No	Suitable
Transesterified Vegetable Oils	No	Good
Hydrogen	Suitable	Conditional

Fig. 5. Applicability of Alternative Fuels

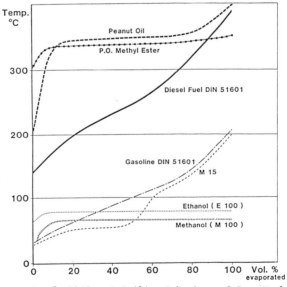

Fig. 6. Different Boiling Behaviour of Some Fuels

	Unit	Gasoline (premium) C_7H_{15}	M100 (Winter) CH_3OH	Ethanol C_2H_5OH
C_5H_{12}	%	-	85	-
H_2O	%	-	<0.1	4.4
RON		98	108	108
MON		88	88	88
Boiling Range	°C	30...200	28...65	78
RVP	bar	0.8	0.9	0.3
Heat of Combustion	kJ/l	32950	17122	20310
Spec.Gravity	g/cm³	0.75	0.779	0.790
Heat of Vaporisation	kJ/l	300	770	714
A/F for λ=1	kg/kg	14.9	7.1	9.0

Fig. 9. Properties of Gasoline, M100, and Ethanol

Characteristic Data		Vegetable Oil	Methyl Ester from Vegetable Oil	Diesel Fuel
Density at 20°C	g/ml	0.92	0.88	0.83
Kin. Viscosity at 38°C	mm/sec²	39	4-6	2-4
Filtration Limit	°C	+15	+10	-15
Cetane Number		40	50	50
Calorific Value	MJ/l	33	32	35
Flash Point	°C	200	150	60

Fig. 7. Data of Vegetable Oil and its Derivatives

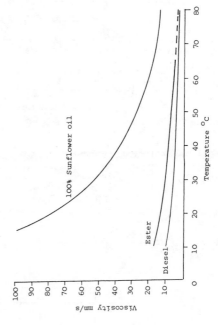

Fig. 8. Viscosity Behaviour of Vegetable Products and Diesel Oil

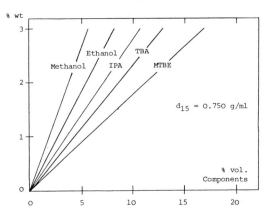

Fig. 10. Oxygen Content of Oxygenated Gasoline

	COMPONENT %			
ALCOHOL	SOLUBILISER	IGNITION IMPR.	DIESEL FUEL
ETHANOL (WATER FREE) 25	5	I	69
METHANOL 15	15	I	69
ETHANOL (WITH 4% WATER) 20	10	1.5	68.5

Fig. 11. Alcohol/Diesel Blend Fuel Data

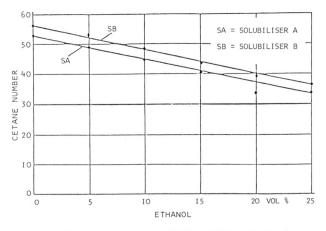

Fig. 12. Cetane Numbers of Ethanol/Diesel Blends

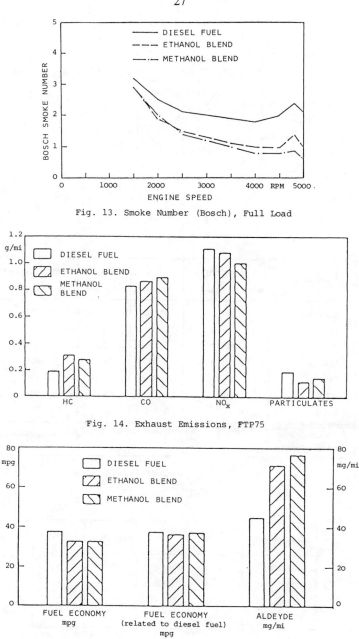

Fig. 13. Smoke Number (Bosch), Full Load

Fig. 14. Exhaust Emissions, FTP75

Fig. 15. Fuel Economy and Aldehyde Emissions, FTP75

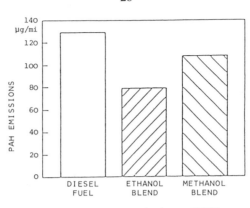

Fig. 16. Total PAH Emissions, FTP75

Component	Oxygen Content Weight % of Molecule	Calorific Value KJ/kg	% of Gasoline	Evaporation Heat KJ/kg
Methanol	50.0	19660	47.6	1100
Ethanol	34.7	26770	64.5	910
IPA	26.6	30060	72.0	700
IBA	21.6	32500	79.6	680
TBA	21.6	33070	81.6	544
MTBE	18.2	35171	79.4	322
ETBE	15.7	36050	81.4	311
For Comparison: Premium Gasoline	–	43500	–	335

Fig. 17. Data of Alcohol and Ethers

Component	RON	MON	Sensitivity	CN
Methanol	114.4	94.6	19.8	3
Ethanol	111.4	94.0	17.4	8
Isopropanol	118.0	101.9	16.1	6
Isobutanol	110.4	90.1	20.6	13

Fig. 18. Octane and Cetane Numbers of Alcohols

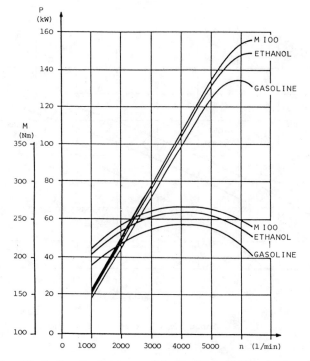

Fig. 19. Engine Performance with Gasoline, M100 and Ethanol

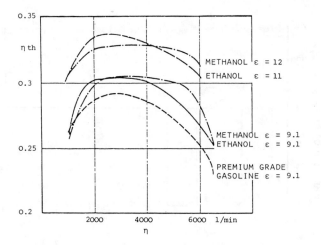

Fig. 20. Thermal Efficiency at Full Load with Various Fuels

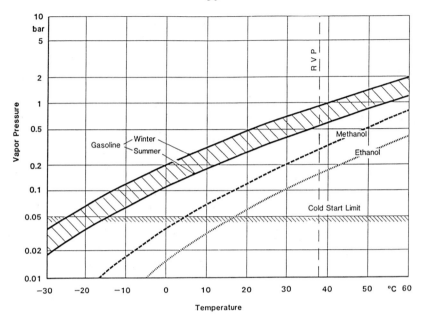

Fig. 21. Vapour Pressure for Various Fuels

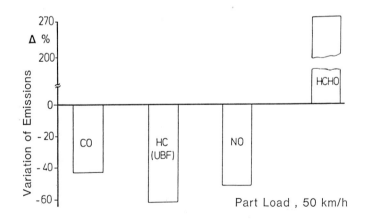

Fig. 22. Emissions of Alcohol Operated Engines Compared to Gasoline

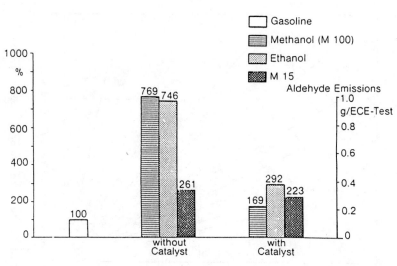

Fig. 23. Aldehyde Emissions with and without Catalyst

Characteristics	Methanol	Ethanol
Thermal Efficiency	+ +	+ +
Cold Start	−	− −
Driveability	0	0
Octane Quality	+ +	+ +
Ignition	−	0
Corrosion , Material Resistance	− −	−

+ +	much better	
+	better	
0	equal	
−	worse	
− −	much worse	

Fig. 24. Methanol and Ethanol in Otto engines Compared to Gasoline

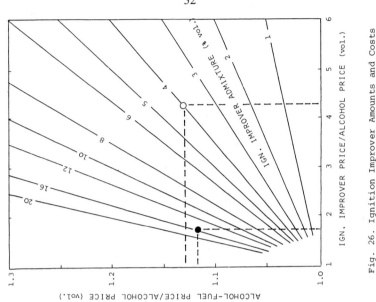

IGN. IMPROVER PRICE/ALCOHOL PRICE (vol.)

ALCOHOL-FUEL PRICE/ALCOHOL PRICE (vol.)

Fig. 26. Ignition Improver Amounts and Costs

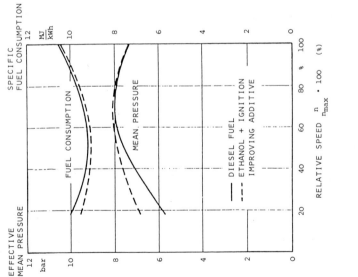

Fig. 25. Engine Performance with Ethanol
and Diesel Oil (Full Load)

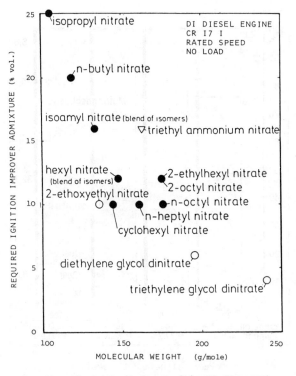

Fig. 27. Products for Ignition Improvement

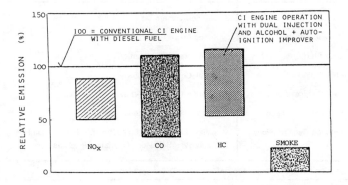

Fig. 28. Emissions with Alcohol in Relation to Diesel Oil

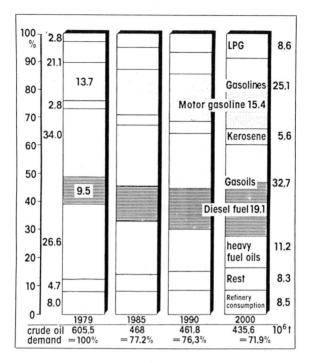

Fig. 29. Crude Oil Demand for Western Europe

Significant Criteria for Gasoline Quality Affected by Oxygenates

Engine Characteristics	Relevant Parameters
Knocking	Octane Numbers *
Driveability	Boiling Curve (Volatility) *
Cold Start	Vapor Pressure , Volatity
Fuel Stability	Water Content , Solubility
Fuel Consumption	Calorific Value , A/F-Ratio
Material Resistance	Fuel Additives , (Material Change)
Exhaust Car Emissions	Ignition Limits (Lean Burn Effects) , Evaporation Heat , Combustion Behavior , Oxydation Products

* Conflict between Octane Rating and Volatility Demand

Fig. 30. Effect of Oxygenates on Fuel and Engine Criteria

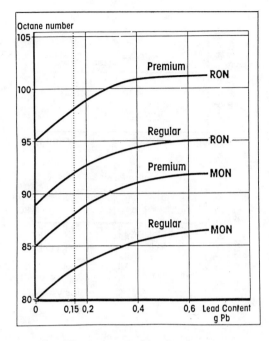

Fig. 31. Octane Quality/Lead Content

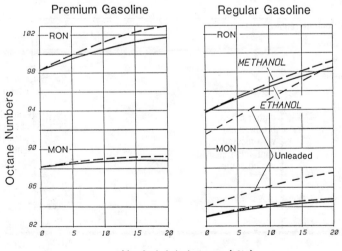

Alcohol Admixture (%)

Fig. 32. Octane Enhancing of Ethanol and Methanol

36

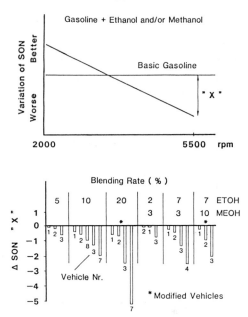

Fig. 33. Variation of Street Octane Number by Oxygenates

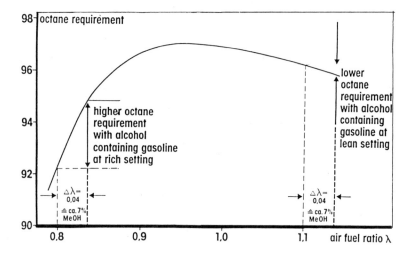

Fig. 34. Octane Requirement versus Air Fuel Ratio

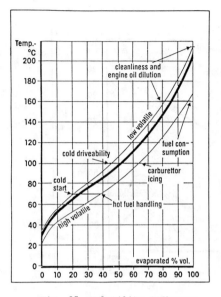

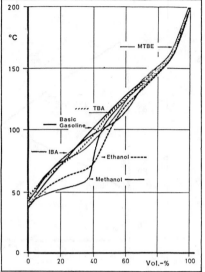

Fig. 35. Volatility Influence on Performance

Fig. 36. Volatility Change by Oxygenates (10% Vol.%)

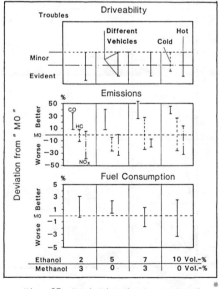

Fig. 37. Variation in Operational Parameters by Oxygenates

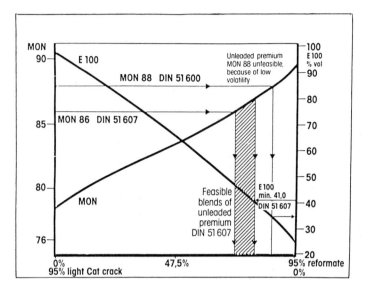

Fig. 38. Conflict between Octane and Volatility Quality

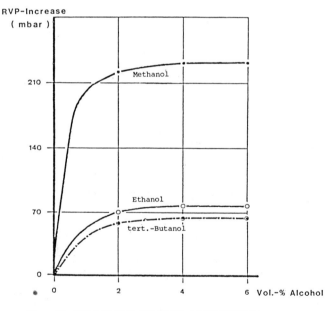

Fig. 39. Reid Vapour Pressure of Oxygenates

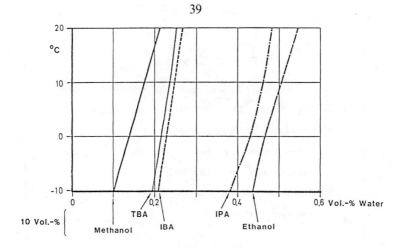

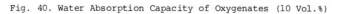

Fig. 40. Water Absorption Capacity of Oxygenates (10 Vol.%)

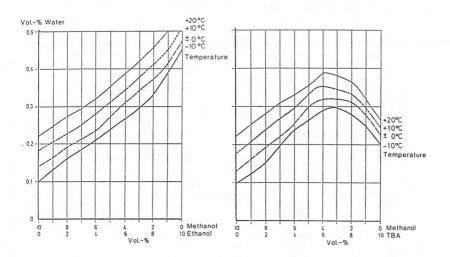

Fig. 41. Phase Separation Temperature for Combined Oxygenates in Gasoline

Project "Eurosuper"

Sponsored by the German Federal Ministry of Food, Agriculture and Forestry (BML)

Fleet Test with a Gasoline – Ethanol – Methanol – Blend Fuel (Eurosuper)

- 10 Vehicles under Operation Control

- 1 Service Station at the BML

- Measurements of Fuel Consumption and

  Exhaust Gas Emissions on Chassis Dynamometer

- Supplementary Investigations

Eurosuper Meets the German Standard DIN 51600

Fig. 42. Project Data 'Eurosuper'

Fuel Components	Vol %
Premium Gasoline (Leaded)	79 – 81
Regular Gasoline (")	9 –11
Ethanol (Hydrous , i.e. 4% Water)	~ 5
Methanol	~ 3
TBA	~ 2
Special Additives	≪ 1

Fig. 43. Composition of 'Eurosuper'

INDUSTRIAL PRODUCTS (NON-FUEL) ORIGINATING FROM
AGRICULTURE: POTENTIAL NEW PRODUCTS

D.A. STRINGER
ICI Agricultural Division, Billingham, UK

1. THE AGRICULTURAL INDUSTRY AS A RAW MATERIAL SUPPLIER TO THE CHEMICAL INDUSTRY

Before considering new products that may be produced by the chemical industry from bio-renewable resources it is sensible to review those that are already manufactured from such products. The major ingredients employed are:

        fats and oils
        starch
        sugar

1.1 Fats and Oils

Fats and oils do not have the same uses as starch and sugar, and need to be considered totally separately. Within the Community 3 m te/y of fat and oils are processed by the chemical industry; most of the raw material is of tropical origin and hence imported. The total may be sub-divided into:

        1.7 mte of human-edible oil (palm, coconut, palm kernel, soya)
        1.3 mte inedible (linseed, tallow, castor)

The chemical industry uses the lower quality and cheaper oils by using inedible tallow from the meat processing industry and acid oils from edible oil refining. The oils produced within the EEC tend to be glycerides with a carbon chain length of C16 or longer, whilst the fats with a mid-chain carbon length, namely C10-C14, derived from palm and coconut, are of tropical origin.

The major manufactured chemical products derived from fats and oils are:

        - alkyds   100 Kte
        - oleochemicals   900 Kte
        - soap   400 Kte
                        ‾‾‾‾‾‾‾
                          1.4 Kte

More specifically, glycerides, fatty acids, methyl esters of fatty acids, fatty alcohols, and all their derivatives.

The uses of fats and fat derived products are numerous, for example,

-       detergents industry:
            washing active substances, emulsifiers, anti-foaming agents
-       paint industry:
            alkyd resins, thixotropic agents, pigment dispersers, epoxy-plasticisers
-       lubricating oil additive industry:
            emulsifiers, anti-corrosion agents, extreme pressure additives
-       plastic additives industry:
            lubricants, anti-oxidants, stabilisers
-       pharmaceuticals industry:
            ointment bases, oil components, suppository bases
-       cosmetics industry:
            cream bases, washing active substances for shampoos, etc. cationic surfactants for hair treatment

- food industry:
        emulsifiers, preservatives, anti-foaming agents
        The chemical industry would welcome the possibility of being less
dependent on the edible oil industry. It has specific needs for various
types of oil/fat, such as:
- short and medium chain fatty acids
- products with a higher content of oleic acid (C16:0)
- unsaturated fatty acids with specific double or triple bondings
- functional groups such as hydroxy or epoxy groups within the fatty
    acid structures
        It is thought that the following plants, which may be grown under the
climate conditions existing within the Community could contribute to the
chemical industry's needs:
        Jojoba, a semi-arid perennial shrub producing a liquid wax ester
incorporating unsaturated alcohols and acids of C20-22 chain length.
        Crambe, an annual plant high in a mono-unsaturated acid of C22 chain
length.
        Lequerella, an annual or perennial herb producing seeds rich in C20
mono-unsaturated fatty acids.
        Cuphea, an annual plant producing short/medium chain fatty acids,
C8-10.
        Euphorbia, a semi-arid perennial plant high in oleic acid (C16:0)
content.
        Sunflower, where it is thought that strain improvement could increase
the oleic acid concentration of the seed from the current 50/60% to greater
than 90%.
        Veronia and Stokesia, which are annual plants with seeds rich in epoxy
oleic acid.
        The chemical industry would welcome R&D activities to demonstrate that
viable crops based on the above may be developed, and that the materials so
harvested may be used by our industry.

1.2 Starch
        Total starch usage within the Community is of the order 3.7 mte/year.
The chemical industry ranks as the third largest user, but only takes 9.5%
of the total output. The major users are the food industry (51%), and the
paper and board industry (20%). Both the chemical and food industries are
major users of hydrolysed starch (80-85% of the total starch) within each
industry total (Table 1).

Table 1 Starch Usage Within the EEC (mte/y)

Total EEC Starch - 3.7 mte/y

	Food	Paper & Board	Chemical
Total	1.88 (51%)	0.74 (20%)	0.35 (9.5%)
Native	0.28 (7.5%)	0.45 (12%)	0.03 (0.8%)
Modified	0.10 (2.7%)	0.29 (8%)	0.02 (0.5%)
Hydrolysed	1.50 (41%)	-	0.30 (8%)

        Native starches derived from either maize, wheat, potato or rice
differ in the physical characteristics, specifically granule size and this
affects end use.
        Native starch is subsequently treated to produce:
- degradation products, specifically oxidised starch and dextrins, which
    are used in the paper and board industries;

- starch derivatives, including starch esters and ethers which are limited in market outlet;
- hydrolysis products, specifically maltodextrins, glucose syrups, and various sugars and alcohols, e.g. dextrose, sorbitol, fructose syrup, levulose and mannitol. Of these dextrose is the most important, in tonnage terms to the chemical industry.

Dextrose (ex starch), and sucrose (ex sugar beet and cane) industries compete as substrates for the fermentation industry.

The application of the Common Agricultural Policy has resulted in starch (and its derivatives) being more highly priced than sources outside the Community. To attempt to overcome this gap the concept of a production refund exists which in theory reduced the price of starch down to world prices. This now is not the case, and the chemical industry is forced to pay prices greater than those prevailing outside the Community. Revisions to the production refund system have been proposed by the Commission so that the production refund will only be paid for industrial uses. How this will affect the price the chemical industry will pay for its raw materials time will tell.

1.3 Sugar

Currently only 62 Kte/y are used by the chemical industry, and the amount is declining.

The principal products for which sugar is used are given in Table 2:

Table 2 Sugar Consumption for Major Chemical Products 1978/79 and 1981/82

	In year 1978/79	In year 1981/82
Total	82.3 Kte/y	62.1 te/y
of which		
Penicillin/antibiotics/ vitamins/amino acids	31.1	16.6
Citric acid	4.0	-
Lactic/tartaric/carboxylic acids	12.6	11.5
Mannitol/sorbitol	12.7	9.0
Levulose/sorbose	5.6	10.8
Polyalkyglycolethers	7.9	8.9
Others	7.4	5.6

Thus production of chemicals by fermentation into those sectors which are not protected by the CAP has been substantially reduced.

As with the starch regime there was the concept that sugar should be available to the chemical industry at 'world prices'. Thus a chemical industry refund is applied to chemicals exported, and internally to those that are not CAP related. At the beginning of the sugar regimes when world sugar was circa £25/te, the refund £25/te, and EEC sugar at £50/te everything was satisfactory. This is no longer the case with 'A' quota sugar at £325/te (540 ECU/te), 'B' quota at £200 (328 ECU/te) the production refund at £33/te (55 ECU), spot 'world prices' at Rotterdam circa £90-100/te, and molasses at £70/te (= £140/te dry solids). Due to EEC CAP regulations the chemical industry cannot import and use free world sugar for products to be sold within the Community. Its alternatives are to use molasses, which can be imported without restraint, or 'A' quota sugar from which it can claim a chemical industry refund for the manufacture of certain products.

Against the background described above of escalating EEC sugar prices it is not surprising that the amount of sugar used by the European chemical industry is declining. The point is well illustrated by the production costs for citric acid (Table 3).

Table 3 Selling & Production Prices for Citric Acid Based on Molasses at £70/te, or Sugar from £100-350/te

Selling price   £960/te

Production cost for a 10 000 te/y plant

		Sugar Cost		
	£70	£100	£200	£350
Variable costs				
molasses	238	-	-	-
sugar	-	170	340	595
other	134	134	134	134
services	39	39	39	39
Fixed costs	192	192	192	192
Return on capital	160	160	160	160
Selling expenses	106	106	106	106
Total cost	869	801	971	1 226
'Profit'	+91	+168	-11	-266

It is therefore not surprising to find that citric acid is no longer produced in this Community from sugar as such. Imported molasses has replaced European sugar.

The important features to be drawn from this situation are:

1. The chemical industry is operating within a true world economic situation. In contrast the European Agricultural Policy is operating in an artificial sector.

2. By allowing agricultural prices to rise without consideration to non-CAP protected products, a separation of the two industries has occurred to the detriment of both. The agricultural industry has lost the market for its products which the chemical industry regards as bio-renewable raw materials.

3. Running parallel with the above discontinuity is the export of Community grown agricultural crops to non-EEC countries, the cost of which could equally well be used to allow the chemical industry to buy raw materials within a European context.

It follows from the above that if the chemical industry is to contribute to a reduction in agricultural surpluses the first condition to be met is adjustment of raw material prices into industrial processes.

2. NEW PRODUCTS FROM THE CHEMICAL INDUSTRY

With the economic structure described above no major developments will take place in the Community that will use sugar or starch hydrolysates. That does not mean that the chemical industry has not been innovative, but that it is seeking investment opportunities for its new plants, processes and products outside the Community.

2.1 Polyhydroxybutyrate

The viewpoint is illustrated by the example of polyhydroxybutyrate, which is a polymer synthesised by bacteria utilising sugar in continuous fermentation conditions. PHB is a new product, it is plastic like, can be made to be biodegradable, but can be made to substitute for polyester or polypropylene plastics manufactured from fossil fuels.

The potential production costs are given in Table 4.

Table 4 Selling and Production Costs for PHB as Influenced by Realisations and Substrate Costs

	Plant Size te/y								
Selling Price: A - medical applications					£20 000 te		10		
B - polyester replacement					£1 400		10 K		
C - polypropylene replacement					£700		50 K		
	100			200			350		
	A	B	C	A	B	C	A	B	C
Variable costs									
sugar	300	300	300	–	600	600	1 050	1 050	1 050
other RNs	162	162	130	–	162	130	162	162	130
services	200	142	70	–	142	70	200	142	70
Fixed	500	55	25	–	55	25	500	55	25
Return on capital	10 000	100	70	–	100	70	10 000	100	70
Total	11 162	750	595	–	1 059	895	11 912	1 509	1 345
Notional profit/te	8 838	650	105	–	341	-195	8 088	-109	-645
Sugar cost as £ of total	2.7	39	50	–	57	67	8.8	69	78

PHB with medical applications, e.g. release control agents for drug therapy, colostomy bags, tampons, selling at about £20 K/te the production cost for a 10 te/y plant is of the order £11-12 K/te to which the price of sugar as the substrate at either £100 or £350/te is largely immaterial. Similarly the amount of sugar consumed, 30 te/y, will not make even a minute impression upon the Community surpluses (circa 3 mte/y).

Larger factories, manufacturing say 10 000 te/y, with product selling price of about £1 400/te (the selling price of polyester) are totally uneconomic with sugar at £350, may break even or make a small reward with sugar at £200, and will be financially successful with sugar at £100/te. With a sugar usage of 3 te per te PHB then 30 K/te sugar would be used, and at this level there begins to be a small impact upon the sugar surpluses.

The effect is much more dramatic upon sugar surpluses if larger chemical plants are constructed, with capacity of say 50 K te/y. Outputs of this variety would outstrip reasonable penetration into the polyester market, leaving polypropylene as the competitive material. The latter sells at £700/te. Under these circumstances sugar at £200 te is too expensive, but when priced at £100 is economic, and what is more 150 Kte/y of sugar would be consumed.

Clearly the economic control of the CAP is having serious disadvantageous effects upon developments in this sector.

2.2 Single Cell Protein

Yeast, produced by the fermentation of sugar, theoretically could replace soyabean meal as a component of animal diets. The current price of soyabean meal ex Rotterdam is £130/te. Historic data, corrected to 1984 prices would indicate that the true price should be £250/te. Even with the latter price it is difficult to justify the case for yeast fodder ex-sugar or starch unless at the same time it is coupled with CAP support provided by savings in foreign currency used for soya purchase. The production costs for yeast are given in Table 5.

Table 5   Realisations and Production Costs for Yeast

	Soya	Yeast (1.3 x soya)		
Current price	£130/te	£170/te		
Potential value	£250/te	£325/te		
Production cost for 10 000 te/y				
	£70	£100	£200	£350
Variable costs				
molasses	198	-	-	-
sugar	-	147	294	515
others	159	159	159	159
Fixed costs	83	83	83	83
Return on capital	83	83	83	83
Total	523	472	619	840

Thus with sugar as low as £100/te the production of yeast is not economic relative to imported soya. The current import cost of soya is £1.6 b/y. By doubling European sugar beet output from approximately 14 m to 28 mte/y sufficient sugar would be produced to provide a fermentation substrate for 10 mte yeast (= 13 mte soya) thereby creating an internal industry with a value of about £4.7 b/y.

2.3 Ethanol
    The above argument is impossible to sustain without major political change. It is difficult to envisage how large-scale agricultural fuel ethanol can ever be regarded as financially sustainable when the market target is an alternative anti-knock component in gasoline. The market value of ethanol/methanol in this sector is £170/te. The production cost of agriculturally derived ethanol is considerably above this value as shown in Table 6.

Table 6 Production Costs for Ethanol ex Sugar Juice (at £106/te)
         and Molasses (£166/te)

Plant size	44K te/y sugar at £106	32K te/y molasses at £166
Variable costs		
sugar	210	370
other RMs	63	101
Fixed costs	41	24
Return on capital	46	52
	360	547

Thus it appears very doubtful whether a case can be made for large scale agricultural ethanol as long as the world gasoline market remains outside the EEC.

3. CONCLUSIONS

1.   Agricultural materials produced within the CAP are too expensive for industrial use.

2.   Structural reform of the CAP is required to allow continued and regular access by industrial users to agricultural materials.

3.   Production of chemicals based on agricultural raw materials is declining within the EEC. A complementary increase is seen outside the Community.

4.   Without reform of the CAP relative to industrial users it is doubtful whether the benefits of an expanded R&D policy will be obtained within the Community.

PRODUCTS DERIVED FROM SUGARS

A. GASET and M. DELMAS
Institut National Polytechnique de Toulouse, France

INTRODUCTION

After water, sugars are the most abundant pure products on the surface of the earth. Since the beginning of time, their primary use has been limited to the nourishment of living beings. Up until the last few years, the parallel growth of sugar production and its consumption as food has considerably limited the availability of these types of resources for other uses and more particularly as a source of raw material in the chemical industry.

There are two phenomena affecting the end of this century which contest this existing 'order of things'.
- The continual and inescapable rise in the prices of crude oil which is linked to geopolitical constraints and to the growing difficulties of its exploitation necessitating a periodical re-evaluation of other sources of carbon substances.
- An agricultural surplus in Europe which is not compensated by an increase in consumption in developed countries due, among other things, to the changes in our food habits which tend to reduce the quantity given to energetic sugar products.

Because of the high amount of oxygen and the sophisticated structures of sugars, sucro-chemistry will have to depend on procedures which will favour techniques able to exploit this incredible potential, thus establishing a close tie between chemistry and biotechnology.

These observations alone, along with the difference in tonnage involved, cancels the opposition too often cited concerning petrochemistry.

In fact, the extreme difference in the raw material has participated in the opening of the market to a very diverse range of products with added value not accessible by petrochemistry and therefore complementary to those produced by industrial organic chemistry.

Based on these elements, our presentation will be focused on chemical or biological reactions liable to open up new channels in the productions of small and medium tonnage, excluding all stages of protection of one or more hydroxyl functions of sugar. The products derived from starch being discussed elsewhere, our paper will focus on the direct transformations already existing and in development for uses other than energetics (ethanol) of glucose, sucrose, lactose, some hexoses available in sizable tonnage and of the main pentoses.

Thus it is with the present situation in mind that this analysis proposes to reflect upon the different possibilities for the development of industrial sucrochemistry by the stimulation of new research in fields which will, on a short and long term basis, open new options for industrial productions of sizable tonnage.

The dominant sugar in the agricultural surplus of the CEE which is the focal point of this operation, is the D-glucose that one finds in sugar plants (saccharose, raffinose, ...), lactoserum (lactose), cereals (starch), wood (cellulose) and in the effluents and other residues of the agro-food industries.

GLUCOSE

The chemical and biological transformations of D-glucose and their potential development will therefore be the main subject of our paper starting with the chemical aspect.

The oxidation of glucose (Figure 1) generally concerns the carbons C6 and C1. The direct oxidation of the aldehyde function leads to the D-glucosic acid produced in amounts of 40 000 tons per year in the world. This molecule and its derivatives are widely put to use in industries relating to food, pharmaceuticals, metal cleaners, concrete additives, and detergents because of their biodegradation and chelating properties. Sodium gluconate has applications which are similar to those of the acid. Moreover, these molecules are able to substitute for polyphosphates in their applications thereby reducing their harmful effects on the environment.

The D-gluconic acid treated by an oxidoreduction system leads to the D-arabinose which is able to be isomerised into D-ribose. Other than the D-xylose, another way to have access to the pentoses, starting with the D-glucose, consists in an oxidation in the presence of platinum followed by a reduction of the acid formed.

The hydrogenation of D-glucose into D-sorbitol is one of the most important industrial operations performed on this product. It corresponds to an annual world production approaching 500 000 tons (Figure 2).

Fifty thousand tons per year is used for the production of Vitamin C according to Reichstein's synthesis which we will come back to later on.

It intervenes in the making of numerous polymers.

The dibenzylidene sorbitol obtained by condensation with the benzaldehyde, used as an additive in the polyethylenes, increases transparency and rigidity.

The esters of sorbitol and of sorbitan are used in numerous formulations of surfactants or emulsifying agents.

The double dehydration of sorbitol leads to dianhydrosorbitol or isosorbide. This compound is not well known, however it is capable of a multitude of potential applications.

It is interesting to note that the double tetrahydrofurannic cycle brings to this molecule interesting properties, particularly in the pharmaceutical field. This suggests a growing effort in the perfecting of new ways to synthesise starting with this compound in order to obtain first generation derivatives.

The animation of glucose on the carbon C1 due to the strong reactivity of the aldehyde is brought about in the presence of ammonia and hydrogen under pressure. In numerous utilisations, the 1-amino 1-dioxy-D-glucose can be substituted for ethanolamine (Figure 3).

The cyanidation of glucose is easily carried out and the additive intermediate product, after hydrolysis, leads to glucoheptonic acid which has the same chelating properties as gluconic acid. Its world production is nearly 5 000 tons per year.

The esterification and the etherification of glucose allows one to obtain molecules which can be used for many applications as intermediaries of synthesis and complements in formulation (Figure 4).

The development of these transformations is undergoing a research of new intermediaries obtained from these molecules by techniques using a single step method which could easily be transposed to a large scale. Such procedures should result in an increase in the consumption of glucose and also in the substitution of derivatives originating from petrochemistry in certain fields of application.

The epimerisation of D-glucose is in itself a most interesting reaction but has the drawback of not being total. Its development is thus

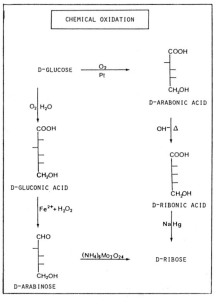

Fig. 1.

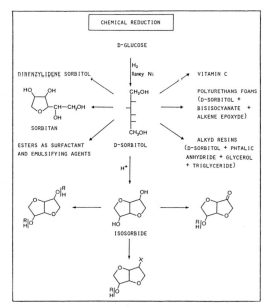

Fig. 2.

51

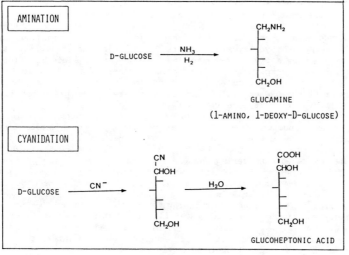

Fig. 3.

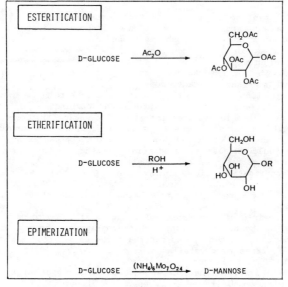

Fig. 4.

hampered by the difficulty in separating the mannose/glucose couple. In any case, we will come back to the general problem involved in the separation of sugars.

A new opening considered by chemists as being very promising, involves the chemistry of hydroxymethylfurfural. However the fragility and the reactivity of this molecule makes its storage and its distribution on an industrial scale complicated. Nevertheless, there seems to be in the world different procedures based on D-fructose coming from the isomerisation of glucose or of the hydrolysis of insulin which are in the pre-pilot stages in certain laboratories (Figure 5).

An important effort must be made to allow a transfer to an industrial scale; an indispensable step allowing for the development and commercialisation of the derivatives of HMF. These must be prepared in situ without previous isolation of the HMF. An entirely new and important market is to be envisaged considering the present prohibitive selling price of HMF.

The strategies of synthesis which are being planned must bring the derivatives of HMF into such a price range as to have them be considered as substrate of classic synthesis.

Following this line of thinking, the reactions of D-glucose with enolisable compounds seem equally capable of creating new possibilities for industry. In fact, the furannic polyols obtained by this reaction have physicochemical characteristics which are particularly interesting for the producers of foams and polyurethanes (Figure 6).

The point here is to bring the cost of production to a competitive level while improving the present methods of synthesis which are adapted to a production at the laboratory level but which remain non-competitive on an industrial scale. Good products to start with would be, for example, molasses, lactose, etc.

Very often interdependent on chemistry and alongside chemistry, biotechnology occupies a decisive position in the industrial transformations of α-D-glucose for the production of molecules of high additional value (Figure 7).

The microbial or enzymatic oxidation participate in:
- an important part of the production of gluconic acid and of gluconates the values of which have been mentioned earlier,
- as well as in the production of the D-isomer of vitamin C via α-ketogluconic acid.

Iso-vitamin C is used at the present time in the amounts of 4 to 5 000 tons per year as an anti-oxidant food.

The production of vitamin C based on D-sorbitol via L-sorbose is a perfect illustration of what the intelligent combination of chemical and biological operations in the transformations of sugars can bring. The value of this example increases in the light of production of ascorbic acid which is based on this well known synthesis and which approaches 50 000 tons per year. This would suggest a substantial backing in the carrying out of research operations based on this procedure.

The isomerisation of D-glucose into D-fructose by the isomerase glucose has a considerable economic importance since more than 4 million tons of isoglucose syrup are already being produced each year according to this technique.

It is possible with an amount of fructose of 42%, to separate it from the glucose by continuous liquid chromatography. The concentration of D-fructose obtained in the syrups can reach 95% to 97% from which it is easy to obtain D-fructose by crystallisation on a yearly production basis which surpasses 20 000 tons (Figure 8).

At this stage, the development (to be encouraged) of hydroxymethyl-furfural chemistry and its derivatives on an industrial scale offers very

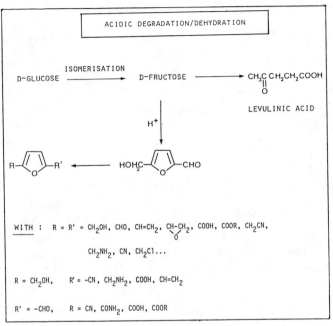

Fig. 5.

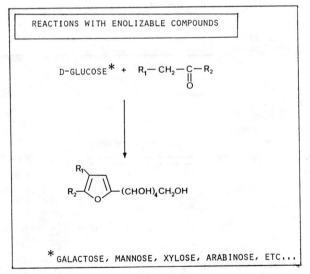

Fig. 6.

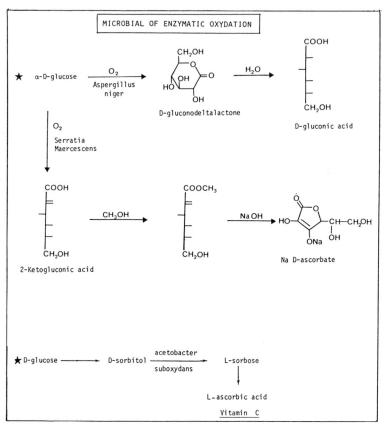

Fig. 7.

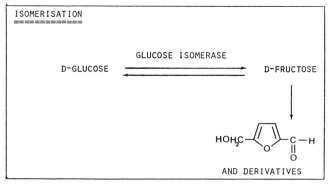

Fig 8.

interesting potential points of development capable of bringing about a measurable increase in the consumption of D-fructose and therefore of its production.

Along these same lines, α D-glucose, by well known means of fermentation (Figure 9) allows one to obtain yeasts, amino acids, antibiotics, alcohols, ketone, organic acids, polyols, etc. Butanediol-2,3 is thus obtained at truly competitive production costs as compared with petro-chemistry. The applications of butanediol-2,3 have brought about an acceleration in the work on the production by biological means of this molecule which is something particularly desirable.

Moreover, the association of chemistry and biotechnology allows one access to the principal food aromas based on D-glucose, such as furaneol (Figure 10).

The biological production of three carbon atom organic acids deserves the following comments:

- An average of 200 000 tons of citric acid is produced in the world and essentially used as acidulaters in food products or in pharmaceutical formulation (Figure 11).
- Itaconic acid, whose production has markedly progressed, is used in the formulation for resins, plastic materials, latex and lubricants.
- Lactic acid is made in amounts of 15 to 20 000 tons per year from glucose, but also from molasses or lactoserum. It is involved in other numerous industrial sectors as well (food industry, pharmacy, fine chemistry).

An increase in the consumption of sugars using these molecules as agents will have to be linked to research of new applications involving the chemistry of their derivatives which should therefore be developed.

Generally speaking, the use of glucose in industrial fermentation (Figure 9) can also contribute to the consumption of sugar surplus by the development, as we have suggested, of new applications, by original means, of derived molecules whose synthesis we can now control.

The biopolymers are, for example, in a competitive position with that of the alginates and other polysaccharides extracts of algae. Nevertheless, the present applications of the dextranes and other xanthanes are still too limited to have any noticeable influence on the reduction of sugar surplus.

Generally speaking, the emphasis must be placed on:

- research combined with the optimisation of production costs in order to go towards a better competition in the main industrial productions that we have mentioned thus far,
- the perfection of new means of finalised transformations of these derivatives. These alone are able to open up new markets which are needed for the present sugar surplus.

SUCROSE

The uses of sucrose for purposes other than food must, of course, be considered as a pure product, but also as being part of the essential sub-products of the molasses industry.

The transformation of sucrose for uses other than food could be considered for different directions which would favour:

- previous hydrolysis in D-glucose and D-fructose;
- the intrinsic reactivity of the hydroxyl sugar groups or its degradation into molecules of low molecular weight;
- the transformation by fermentation which is particularly adapted to the valorisation of molasses.

The D-glucose and D-fructose syrups produced by hydrolysis have the drawback of requiring the separation of the two sugars. It is seemingly

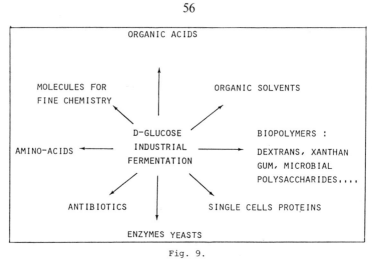

Fig. 9.

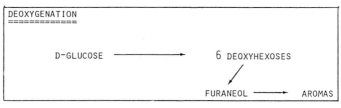

Fig. 10.

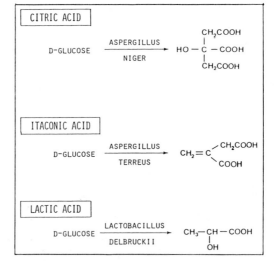

Fig. 11.

easier to do this on a mixture of the two polyols obtained by direct hydrogenation reaction on the sucrose, that is to say, D-sorbitol and D-mannitol (Figure 12).

This means of transformation seems promising considering the present markets for these molecules and in particular of sorbitol, mentioned earlier, as well as the remarkable but little-exploited potential of their derivatives, and notably of the corresponding dianhydrohexitols.

The production of glycerol and of glycol propylene by hydrogenolysis at high temperatures also deserves to be taken into account although it is highly in competition with the petrochemical processes due to the expense of energy necessary for the synthesis of these products.

In the same way, according to the conditions of the operation, oxalic acid and other hydroxyl organic acids can be obtained by the oxidation of sucrose and thus recalls the applications mentioned previously, particularly in what concerns the gluconic acids. Arabonic acid, which can also be obtained in the presence of platinum, is then transformed into D-ribose which participates in the synthesis of vitamin B2.

The chemical transformations of sucrose of industrial interest, grouped together in Figure 13, provoke either an important degradation of the molecule or a reactivity of its hydroxyl functions.

Thus it is that by heating sucrose at high temperatures (200°C) in the presence of ammonia and hydrogen, one obtains methyl-2 piperazine capable of integrating the fabrication of different polyamides. Some aminoalcools and other diamines are formed during this reaction. However, this last one is not selective enough yet to merit an important development.

In a basic medium and at high temperatures (240°C), lactic acid can be obtained with good results (70%) and can therefore be competitive with the traditional productions by fermentation of molasses or of lactoserum previously mentioned. The significant development of such a means of synthesis and therefore of its effect on the consumption of sucrose is directly related to an industrialisation of the chemistry of lactic acid and its derivatives. It will be a question of optimising, for example, the production of products as important as acrylonitrile or methyl acrylate which are direct derivatives (Figure 16).

The degradation in an acid medium leads, according to the conditions of reaction, to hydroxymethylfurfural (with limited results) or to levulinic acid. These two products are remarkable intermediaries of synthesis. The development of the chemistry of their derivatives seems to be extremely promising because of their ability to develop rapidly towards productions of medium tonnage.

The esterification of sucrose is, at the present time, the most significant industrial application (25 to 30 000 tons per year) and seems to be in continual progression (detergents and polymers).

The octoesters, which are the result of the esterification of the eight hydroxyl functions of sucrose by acetic anhydride or benzoyle chloride, find very interesting and specific applications in the fabrication of adhesives, laminated glass, resins, etc.

The fatty acid esters are excellent non-ionic surfactants having the following main advantages:

-    an easy synthesis which is relatively selective of the monoesters at a very low cost;
-    are non-toxic and easily biodegradable;
-    offer a return on capital invested and particularly interesting potential profits on a unit production basis of 5 000 tons per year, which leaves an important opening for industry in years to come.

58

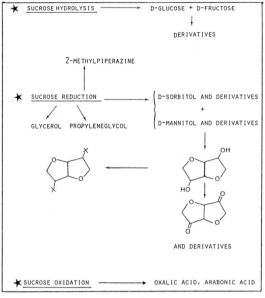

Fig. 12.

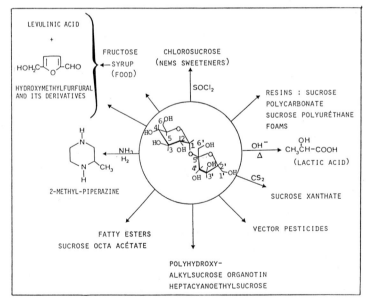

Fig. 13.

The xanthates of sucrose, another type of surfactant derived from sucrose, are obtained by successive reactions with carbon disulfide and an alkyle halogenure, however, they are of lesser importance as of yet.

The etherification of sucrose has known an appreciably smaller development. The only significant applications have come from the cyanoethylation of acrylonitrile which leads to compounds with dielectric properties of the formation of polyhydroxyalkylethers with propylene or ethylene oxide.

The utilisation of sucrose in rigid polyurethane foams is in constant progression thanks to the presence of the eight hydroxyl functions; so it is that for a market of 200 000 tons per year, 20% of the production already integrates sucrose as the basic polyol.

In the same line of thought, but on a smaller scale, one can note the formation of polycarbonates resins based on sucrose and ethyl chloroformate.

Sucrose can equally be a support and even a vector of active principles as in the organostannous pesticides by complexation or formation by covalent bonding. This application, which is of minor importance at the present time, appears most promising and could be a stimulant for an important consumption of sugars since this technique is not specific to sucrose. Thus diverse sugars can take on this position of vector. The choice of the sugar will be related to its cost.

On the other hand, the inevitable development of the chlorized derivatives of sucrose may create supplementary problems in terms of surplus inasmuch as the enormous sweetening power of tetrachlorogalactosucrose should markedly reduce the consumption of sucrose in agro-foods. This underlines the urgency for the development of new industrial openings around this molecule and its present derivatives.

The fermentations must, with chemistry, meet this challenge (Figure 14).

Naturally one rediscovers applications very similar to D-glucose such as with the action of glucose oxidase which produces a mixture of D-fructose and gluconic acid which are both highly valued on the industrial level.

The different proposals and orientations which have been mentioned in what concerns the future of sucrose chemistry constitute only a part of the developments in industrial usage of this substrate as a primary material. This implies a considerable amount of research in the perfection of original techniques; a research which must be strongly encouraged.

LACTOSE

The situation is much the same for lactose, which is capable of being produced in sizable tonnages by means of the recuperation of lactoserum (Figure 15).

There are four essential directions to be developed similar to that of sucrose and according to a parallel process leading towards:
- transformation by fermentation;
- direct functionalisation by means of a study on the specificity brought about by lactose in relationship to sucrose;
- direct chemical transformation of the molecule;
- and, of course, hydrolysis with particular attention paid to the problems caused by the separation of glucose/galactose.

This is, of course, the main obstacle in the valorisation of sugars which are present in minor quantities but which could attain valorisations of considerable interest.

Therefore, the industrial separation of the different sugars such as hydrolysate sugars in wood, deoxygenised sugars of bacterial walls, etc. are too poorly controlled at this time for these different hexoses to become primary industrial products. Their valorisation relies on the perfection of

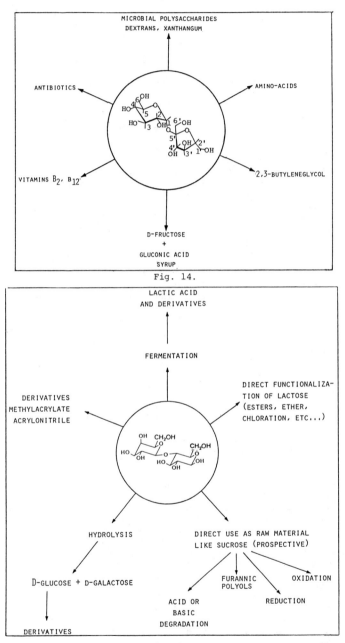

MICROBIAL POLYSACCHARIDES
DEXTRANS, XANTHANGUM

ANTIBIOTICS

AMINO-ACIDS

VITAMINS B_2, B_{12}

2,3-BUTYLENEGLYCOL

D-FRUCTOSE
+
GLUCONIC ACID
SYRUP

Fig. 14.

LACTIC ACID
AND DERIVATIVES

FERMENTATION

DERIVATIVES
METHYLACRYLATE
ACRYLONITRILE

DIRECT FUNCTIONALIZA-
TION OF LACTOSE
(ESTERS, ETHER,
CHLORATION, ETC...)

HYDROLYSIS

DIRECT USE AS RAW MATERIAL
LIKE SUCROSE (PROSPECTIVE)

D-GLUCOSE + D-GALACTOSE

FURANNIC
POLYOLS

OXIDATION

ACID OR
BASIC
DEGRADATION

REDUCTION

DERIVATIVES

Fig. 15.

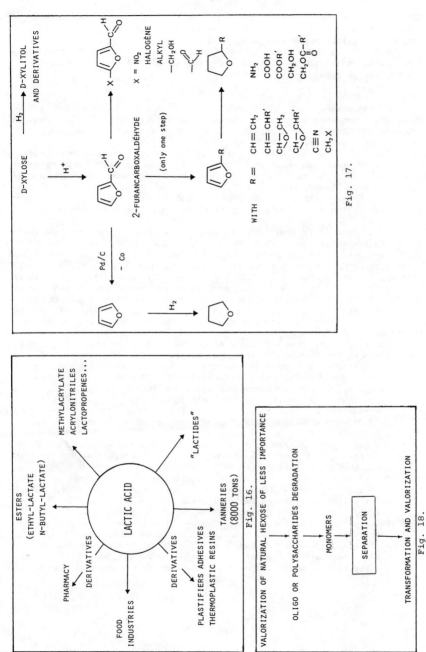

Fig. 17.

Fig. 16.

Fig. 18.

efficient separating techniques which could be industrialised and would combine the techniques of biotechnology and chemistry to which we are today giving our fullest attention (Figure 18).

Within this same perspective, D-mannose and its derivatives could have interesting industrial application for agro-food products, pharmacy, and polymer chemistry.

XYLOSE

If the hexoses and their various forms occupy the major part of sucro-chemistry, one must not neglect the pentoses and more specifically D-xylose produced essentially from sequenced hydrolysis of vegetable residues (corn cobs, sugar cane stems, olive pits).

Its hydrogenation into xylitol has not been developed as expected, however this means of valorisation is enjoying a renewed interest which should hopefully be maintained in light of its numerous applications in agro-food products (Figure 17).

It is essentially the dehydration into furfural (200 000 tons per year) which constitutes the major part of the industrial transformations of D-xylose.

It is especially the fireproof properties of the furannic cycle which are being presently exploited in the resins of the foundry industries and in a few thermo-resistant polymers. This, associated with the appearance of new techniques which allow the transformation of furfural into such basic monomers as 2-vinylfuranne, 2-furyloxiranne, furylacrylates, and their tetrahydrofurannic homologues, suggests an encouragement in the development of directions of research capable of backing an industrialisation of different ways to produce new polymers or thermo-resistant resins (polyurethanes, for example) for which the existing and future markets are important.

In conclusion, the key ideas which appear to us capable of bringing new answers to the problems of the sugar surplus in the EEC are:

- A call for an increased consumption of this surplus by the development of the chemistry of the main derivatives which are basic to sugars (polyols, furfural, dianhydrohexitols, hydroxymethylfurfural, lactic acid) in view of their numerous applications.
- The research of new procedures or an optimisation of present processes of fabrication which would favour productions of medium tonnage at added value.
- The development of the uses of sugars and of their derivatives in the chemistry of polymers towards new formulations associating the main petrochemical intermediaries.
- The favouring of research projects which integrate a basic understanding and consideration of the economic constraints which exist in the present and future markets concerning the availability of raw materials and the products which can valorise them.
- Finally, to favour systematically the mutually completing sciences of chemistry and biotechnology, chemical engineering and biological engineering which are capable of opening up new strategies in the transformations of sugars and their derivatives and thereby contributing to a supplementary consumption of the sugar surplus.

PRODUCTS DERIVED FROM STARCH

H.U. WOELK

Maizena GmbH, Maizena-Haus, 2000 Hamburg 1, Federal Republic of Germany

Starch made from grain or potatoes is today already a versatile raw and auxiliary material that is widely used even outside the food sector.

Almost 1.5 million tons, or more than 40%, of the starch products manufactured in Europe go into different non-food industries, more than 400 000 tons being absorbed by the chemical industry alone.

This by no means exhausts the technochemical potential of the polymeric carbohydrate. New possibilities for a more intensive utilisation of these annually renewable resources by the chemical industry outside the field of biotechnology appear to be particularly interesting because they could open up very large markets medium term. The exploitation of this potential would help solve some pressing agropolitical problems, especially that of a meaningful use of agricultural surpluses.

Following a basic analysis of this potential, it is shown by some examples how starch products used in plastics production or plastics processing are capable of changing the property profile of these plastics, e.g. their mechanical properties, their behaviour in fire, or their environmental compatibility. Thanks to their specific structures and reactivities, starch-derived products may also be important building blocks in the chemical synthesis of, for example, tensides or pharmaceuticals.

A first rough estimate of the sales potential to be opened up in this way for starch products leads to a figure of over 1.6 million tons, which is four times the volume presently consumed by the chemical industry.

To produce this quantity, an area of 1.66 million acres would be needed if this starch were obtained exclusively from wheat.

However, there are two prerequisites for a commercial exploitation of these and other newly emerging possibilities of starch as a renewable raw material:

- much more intensive chemical and especially interdisciplinary research into the various aspects,
- the removal of still existing economic disadvantages caused by agropolitical regulations relating to the industrial use of agricultural resources,

which are equally discussed by reference of several examples. Unless a favourable agropolitical environment is created very soon, it will be impossible to prevent the exploitation of the starch potential as a renewable raw material taking place chiefly in non-EEC countries, or being shifted into these countries.

PRODUCTS DERIVED FROM HEMICELLULOSES AND LIGNIN

H.H. NIMZ
Institute for Wood Chemistry and Chemical Technology of Wood
Federal Research Centre for Forestry and Forest Products
Leuschnerstrasse 91, 2050 Hamburg 80, Federal Republic of Germany

OCCURRENCE AND AVAILABILITY OF HEMICELLULOSES AND LIGNIN

After cellulose, lignin and hemicelluloses constitute the most abundant organic renewable raw materials. According to Lieth (1973) the annual growth rate for both of them worldwide amounts to some 12 billion tons. In the pulp mills both are obtained as by-products in the spent liquors, which are mainly burnt. In this case they serve as energy suppliers for the pulp mills. However, this kind of utilisation is not very satisfactory for the following reasons:

-   Vast amounts of water have to be evaporated from the aqueous solutions before burning them, so that most of the energy gained is already needed as evaporation energy.
-   Due to the high amounts of chemicals in the spent liquors, SO_2, mercaptans and other acidic, toxic or corrosive gases are formed, which for environmental reasons have to be absorbed from the off-gases carefully before releasing them into the air, causing additional costs.
-   Lignins as well as hemicelluloses, unlike coal or crude oil, constitute valuable organic raw materials with specific chemical structures, namely polyphenols and polysaccharides.

From the pulp mills, some 50 million tons of lignin and 30 to 40 million tons of hemicelluloses are produced annually worldwide. For comparison, the global production of plastics amounts to some 40 million tons.

Another powerful source for hemicelluloses and lignin arises from agricultural by-products, namely cereal straw. According to Baudet (1984), the global production of cereals in 1981 amounted to 1.2 billion tons.

Table 1 Lignin and Hemicelluloses as By-Products of Chemical Pulp and Cereal Crop in the EEC 1982, in 1 000 t

		Chemical pulp	Lignin	Hemicelluloses
Wood [1]	prod.	3 091	1 500	1 000
	cons.	7 645	120*	80*
Cereal straw [2]	prod.	200	100	150
140 000	potent.	50 000	22 500	42 000

1 VDP, 1984
2 F. Rexen and L. Munck, 1984
* Estimated values

Table 1 shows that the production of chemical pulp in the EEC amounted to some 3 mio tons in 1982, giving rise to some 1.5 mio tons of lignin and 1 mio tons of hemicelluloses in the spent liquors. The latter are mainly burnt and only some 20% are spray-dried and sold for different purposes (see below). Table 1 also reveals that only about 40% of the consumed chemical pulp is produced in the EEC, while the residual 60%, namely 4.5 mio tons, have to be imported mainly from North America and Sweden. This situation creates the urgent need for other pulp raw materials than wood, which possibly might be cereal straw, of which the EEC has a high surplus. The present production of chemical pulp from straw makes up only a minor part of this from wood. The main reasons are transport and other logistic problems, allowing only small-sized straw cellulose factories that cannot compete with the large North American and Swedish pulp mills.

Only about 10% of the potentially available straw would cover the presently imported chemical pulp in the EEC, which is a feasible amount taking into account other utilisations such as cattle litter and fertiliser for soils, as well as transport problems and physical problems of straw fibres. These problems have been discussed in more detail by Rexen and Munck recently (1984).

Table 2 Content of Main Constituents in Wood and Straw (%)

| | Cellulose | Lignin | Hemicelluloses | | Ash |
			Xylans	Glucomannans	
Softwood	41-43	27-33	10-15	15-25	0.3
Hardwood	41-43	20-24	10-35	3-5	0.4
Wheat straw [1]	37	16	25	3	4
Oat straw [2]	37	15.4	25	n.d.	2.6
Oat hulls [1]	33	10	12	2	4
Rye straw [2]	37	17.6	26	n.d.	1.2
Rye internodes [2]	40	18	25	n.d.	0.4
Barley straw [3]	37	6	25	n.d.	6

1 Puls and Hock, 1985, recalculated from total sugar contents

2 Muller, 1960

3 Puls, 1983, recalculated from total sugar contents

Table 2 indicates the relative amounts of lignin and hemicelluloses in wood and straw. It can be seen that softwoods contain more lignin and less hemicelluloses than do hardwoods while the content of cellulose is nearly the same for European wood species. There is another characteristic difference between hardwood and softwood hemicelluloses in that the former consists predominantly of xylans (pentosans) while in the latter the glucomannans (hexosans) predominate. Cereal straws are similar in their composition to hardwoods with only small but characteristic differences: they contain less cellulose and lignin besides significant amounts of silica, which may cause problems in straw pulping. There also exist differences in the composition of nodes, internodes, leaves and hulls, as can be seen from Table 2. Table 2 also reveals that straw and especially oat hulls are rich in xylans giving rise to high yields of furfural (see below) when treated with mineral acids.

COMPOSITION OF HEMICELLULOSES

Table 3 Main Sugars Obtained from Three Typical Hemicelluloses
According to Puls and Hock (1985) (%)

Hemicelluloses of	Glu	Xyl	GluA	Ara	Gal	Acetic acid
Beech wood	1.4	70.8	13.7	0.1	-	4.35
Wheat straw	2.1	70.8	5.5	10.3	1.6	2.7
Oat hulls	-	75.8	2.3	10.2	1.7	7.3

The composition of hemicelluloses may be seen best from Table 3, which lists the sugar composition of three typical hardwood and straw hemicelluloses after total hydrolysis by different methods. Hemicelluloses of hardwoods and cereal straws both contain more than 70% xylose, while differing characteristically in their amounts of glucuronic acid and arabinose. Hexoses, like glucose, galactose, and mannose, together make up less than 5%, while in hemicelluloses of softwoods hexoses predominate over pentoses. This difference between hemicelluloses from hardwoods and straw on the one hand and from softwoods on the other hand is of basic importance for their transformation to products like ethanol and furfural, as will be shown later.

CHEMICAL STRUCTURE OF LIGNINS

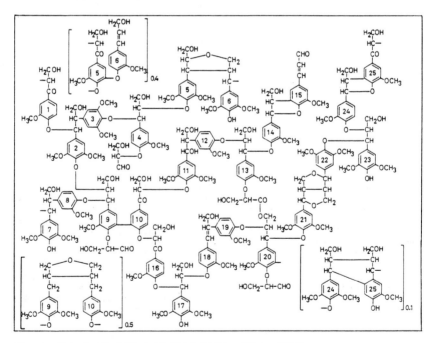

Fig. 1. Structural Scheme of Beech Lignin (Nimz, 1974)

Hemicelluloses differ from lignins in that they consist of slightly branched chains which can be hydrolised to monomeric sugars while lignins constitute cross-linked polyphenols consisting of phenylpropane structural units. Figure 1 shows a structural scheme of beech lignin. It can be seen that the C9 units are linked by carbon-carbon as well as carbon-oxygen bonds, which make the degradation to uniform monomeric phenols impossible. This is one of the main problems for the utilisation of technical lignins. Softwood lignins differ from hardwood lignins in having less methoxyl groups and a higher ratio of carbon-carbon to carbon-oxygen bonds which makes the degradation of softwood lignins even more difficult than that of hardwood lignins.

LIGNIN DEGRADATION PRODUCTS (Figure 2)

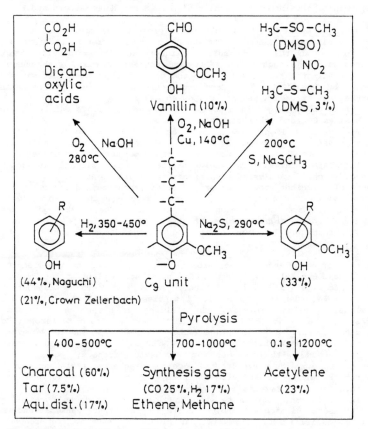

Fig. 2. Products Derived from Lignin by Degradation

Vanillin has been produced in the USA and Canada (Rotschild, Ws, Cornwall, Ontario) since 1938 by oxidation of spent sulfite liquor. Other plants were built after the second world war in Europe (Borregaard, 1963,

200 t/y), in Japan and the USSR. However, the yield of vanillin is low (8-10%, based on lignin) and its price (3 US dollars per pound) is too high for a wide application. It is mainly used as a flavouring and odour masking agent, furthermore as a reductant, UV absorbing material, for instance as a protecting agent against sun burn, as well as for the syntheses of many pharmaceuticals. Vanillin is only obtained from softwood spent sulfite liquor, because hardwoods would give mixtures with syringaldehyde.

DMS is obtained from lignin (sulfite or sulfate) with Na_2S or sulfur at temperatures above 200°C. A plant was built by Crown Zellerbach Corp. in Bogalusa, Louisiana, with a capacity of 5 000 t/y, in 1960. The yield is very low, 3% based on lignin, due to a theoretical yield of only 7-10%. DMS is oxidised by NO_2 to DMSO, which is an excellent solvent and outstanding medium for promoting chemical reactions.

At higher temperatures (290°C), a mixture of low molecular weight phenols can be obtained from lignins with Na_2S and NaOH (Asborn and Enkvist, 1962).

Hydrogenolysis at 350-450°C in the presence of catalysts leads to a mixture of alkylated phenols. The Japanese Noguchi process consists of two steps, by which the spent sulfite liquors are first desulfonated and separated from the sugars and salts and thereafter mixed with phenol in order to render the lignin soluble in the hydrogenation step. It was shown, however, by Crown Zellerbach Corp. that the yields given by the Japanese researchers (44%) were unrealistic (Goheen, 1971).

Oxidation by oxygen in a sodium hydroxyde melt leads mainly to oxalic acid, besides CO_2 and other carboxylic acids.

Pyrolysis of lignin takes place at temperatures above 250°C in the absence or deficit of oxygen. At temperatures below 500°C only small amounts of phenols (as tar), acetic and formic acid, methanol and acetone (aqueous distillate) are obtained. The yields may be increased slightly in vacuum or by the addition of catalysts, anthracene or tetraline. At 700-1 000°C in the presence of deficient amounts of oxygen or air, synthesis gas (CO + H_2 (N_2)) is formed which can be used for the synthesis of methanol. Recently fluid bed techniques have been developed for wood (Nitschke, 1981) gasification which should give higher yields of CO and H_2 with organosolv lignins. Ethene, methane and benzene can also be obtained from lignins between 700 and 1 000°C mainly in the absence of oxygen.

Flash pyrolysis is used in the petrochemical industry for the production of acetylene (ethine) from natural gas. When this procedure was used by the Crown Zellerbach Corp. on a pilot scale plant in the 1970s, up to 23% kraft lignin and 10-15% of finely ground wood could be transferred to acetylene. In order to compete with the price of acetylene, obtained from natural gas by flash pyrolysis, the yield has to be at least 30% and the capacity of the plant 150 t/d. For this reason, Crown Zellerbach Corp. decided not to build a larger plant (Goheen, 1978). It should, however, be mentioned that higher yields are possible with sulfur-free lignin instead of kraft lignin and with other than a combustion furnace, used by Crown Zellerbach Corp., i.e. electric arc, corona discharge, resistance furnace, regenerative furnace or induction furnace, that are all described in the literature for flash pyrolysis of hydrocarbons.

In conclusion, it can be said that the only products obtained from lignin by degradation on a technical scale at present are vanillin and DMSO. The other possibilities, summarised in Figure 2, have not yet reached an economical level. The main reasons are the low yields and high diversity of the products in all cases. The yields, however, could be improved markedly, if the technical lignins were less condensed and free from carbohydrates, inorganic salts and sulfur. The latter particularly poisons the catalysts necessary for hydrogenolysis.

Polymeric Lignin Products from Spent Liquors

In 1982 the EEC produced some 1 100 000 t of solids in spent sulfite liquors and 2.2 mio t black liquor solids in kraft mills (VDP, 1984). The solids of the black liquor consist of up to 40% inorganic salts and they are completely burnt in the EEC in order to recover the pulping chemicals. Only the spent liquors obtained from calcium bisulfite mills are mainly spray-dried and sold for a great variety of purposes (Table 5). Their amounts within the EEC may roughly be estimated to be some 200 000 t/y spray-dried lignin sulfonic acids.

Table 4 Composition of Calcium Base Spent Sulfite Liquor (%)

	Softwood	Hardwood
Lignin sulfonates	55	42
Hexoses	14	5
Pentoses	6	20
Sugar acids and residues	12	20
Resins and extractives	3	3
Ash	10	10

Their composition is shown in Table 4. It can be seen that these powders contain about 10% inorganic salts and between 10 and 45% carbohydrates. The residue (42-65%) consists of lignosulfonates. The hexoses can be fermented to ethanol (up to 100 l per ton of pulp) and the pentoses to yeast (Torula or Pekilo yeast). However, production of ethanol from sulfite liquor has declined since the 1950s and nowadays is no longer the practice in most pulp mills, due to the low price of ethanol from ethylene.

Table 5 Applications of Spray-Dried Spent Sulfite Liquors (SSL)

Binder and adhesive	Pelletising animal fodder
	Substitute to phenolic resins
	Adhesive for particle boards
	Ore and mold binders
Grinding aid and additive for cement	
Dispersants	Clay and ceramic industry
	Paints
	Insecticides
	Herbicides
	Pesticides
Emulsifier and stabiliser for oil-in-water emulsions	
Oil well drilling mud	
Tanning and sequestering agent	

Applications of spray-dried spent sulfite liquors (SSL) can be seen from Table 5. Some 60% of the SSL are used for pelletising animal fodder, 30% as oil well drilling mud, cement and concrete additives, dispersants and emulsifiers. Addition of 0.1-0.3% of SSL to cement allows water reduction by 20% and improves the compressive strength and durability of the concrete. As a dispersant, SSL imparts a negative charge to the particles, hindering their agglomeration in water. It also lowers the viscosity of the dispersion and works as a surfactant. As a dispersant, it imparts proper rheological properties to drilling muds in the oil industry and has replaced tannins, that have been used previously. SSL can also be used instead of tannins for leather tanning. Being a sequestering agent means that SSL can hinder the formation of insoluble hydroxides from metals like iron, aluminium, calcium, zinc and others at high pH values or rendering insoluble salts in solution. SSL is, for instance, used for complexing magnesium and calcium salts for their application on trees avoiding brown spots on the needles and making the salts available to the roots.

Products from Hemicelluloses (Figure 3)

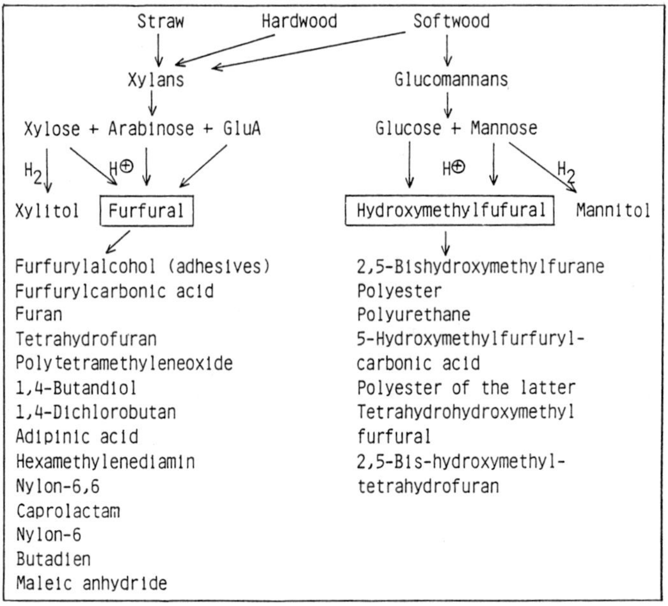

Fig. 3. Main Products Derived from Hemicelluloses

As has been pointed out earlier, straw and hardwood hemicelluloses consist mainly of xylans (pentosans), while softwood hemicelluloses consist of both xylans and glucomannans. Hydrolysis of xylans leads mainly to xylose (cf. Table 3), arabinose and glucuronic acid, while that of glucomannans leads to glucose, mannose and galactose. Besides getting xylitol from xylose and mannitol from mannose, the products that can be obtained in highest yields are furfural from pentoses and hydroxymethylfurfural from hexoses. Both compounds are important chemical raw materials from which a great variety of chemicals and polymers (polyethers, polyamides, polyesters, polyurethanes, polybutadiene) may be obtained. The crucial question here is the cost of the two key products, furfural and hydroxymethylfurfural (Schliephake and Otzen, 1985). Up to now, most of the products obtained from furfural listed in Figure 3 can be obtained at a lower price from petrochemicals. The present price of furfural is based on its production from oat hulls or maize straw and depends on the concentration of xylose in the solution from which furfural is obtained, which in this case is about 30%. It is less in SSL from hardwoods (cf. Table 4). However, if the hemicelluloses in the SSL from hardwoods could be separated from the lignin and the salts, the xylose concentration would increase to 50% or even higher. The same would be true, if the hemicelluloses could be isolated from cereal straws, yielding up to 70% xylose.

Acetosolv Pulping (Figure 4)

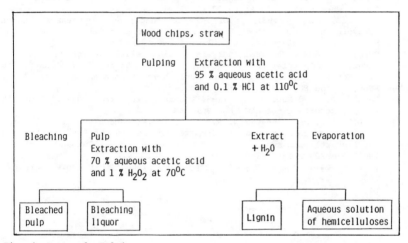

Fig. 4. Acetosolv Pulping

We have recently developed a pulping procedure by which all three wood components, cellulose, hemicelluloses, and lignin, can be separated under mild conditions (Figure 4). Wood chips or straw are extracted with 95% aqueous acetic acid containing 0.1% hydrogen chloride at 110°C for four hours. The pulp may subsequently be bleached by adding 1% hydrogen peroxide to the acetic acid, forming a very selective oxidant for lignin, peracetic acid. On pouring the concentrated extract from the pulping step into water, the precipitated lignin can be filtered off, and an aqueous solution of pure hemicelluloses is obtained. In this way we obtain a low condensed lignin, free from sulfur, inorganic salts and carbohydrates, which should be more convenient for hydrogenation to phenols, pyrolysis to synthesis gas and acetylene. On the other hand, the hemicelluloses are free from lignin and inorganic salts. In the case of hardwood and cereal straw they should be rich in pentoses, thus reducing the price for furfural.

The new acetosolv process does not lead to any sulfur or chlorine containing organic chemicals or inorganic salts. The amount of wash waters is much less than in conventional pulping and bleaching procedures.

REFERENCES

(1)   ASBORN, T. and ENKVIST, T. (1962). Proofs for the occurrence of homoprotacatechuic acid and some other phrochatecol and guaiacol derivatives in pressure heated cellulose spent liquors, Acta Chem. Scand. 16, 548-552.

(2)   BAUDET, B. (1984). The Agricultural Trade War, Bureau d'Informations Européennes, SPRL, Brussels, Belgium.

(3)   GOHEEN, D.W. (1971). Low molecular weight chemicals, in: Lignins, K.V. Sarkanen und C.H. Ludwig (Ed.), Wiley-Interscience, New York.

(4)   GOHEEN, D.W. (1978). The preparation of unsaturated hydrocarbons from lignocellulose materials, Cellulose Chem. Technol. $\underline{12}$, 363-372.

(5)   LIETH, H. (1973). Human Ecology, $\underline{1}$, 303.

(6)   MULLER, F.M. (1960). On the relationship between properties of strawpulp and properties of straw, Tappi 43 (2), 209-218.

(7)   NIMZ, H. (1974). Beech lignin - proposal of a constitutional scheme, Angew. Chem. internat. Edit. $\underline{13}$, 313-321.

(8)   NITSCHKE, E. (1981). Die Herstellung von Synthesegas durch Holzvergasung nach dem Rheinbraun-HT-Winklerverfahren. Ber. d. Kernforschungsanlage Jülich. Jül-Conf. $\underline{46}$, 231-246.

(9)   PULS, J. (1983). Chemical analysis of lignocellulosic residues, in: A. Strub, P. Chartier and G. Schleser (Ed.), Energy from biomass, 2nd E.C. Conference, Applied Science Publishers, London and New York 1983.

(10) PULS, J. and HOCK, R. (1985). Unpublished results.

(11) REXEN, F. and MUNCK, L. (1984). Cereal crops for industrial use in Europe, Report of the CEC, EUR 9617 EN.

(12) SCHLIEPHAKE, D. and OTZEN, P. (1985). Marktchancen von Produkten aus Biomasse im Energie- und Chemiemarkt mit besonderer Berücksichtigung der Produktlinien aus Oelen und Fetten sowie Lignocellulose, Teil II. BML-215-79/2, Proj.-Nr. 82 NR 005.

(13) VDP. (1984). Annual Report of the Verband Deutscher Papierfabriken, FRG.

FUEL PRODUCTS FROM BIOMASS

R. GABELLIERI

Centre d'Etudes des Communautés Européennes, Brussels, Belgium

The production of alternative fuels from biomass represents a good opportunity for EEC agriculture: development of already existing crops, introduction of new cultures, better utilisation of farming by-products, etc.

The first question to answer is: has EEC agriculture the possibility to cope with a large demand of ethanol, the alternative fuel usually considered as the easiest to obtain from biomass?

The answer is positive. This potential does exist, staggered in three steps:

1. Existing farming structures can satisfy up to a demand of about 10 million tons of ethanol.

2. This potential can be roughly doubled by the use of new crops, already known but not yet spread in the complex world of the farming activity of the Community.

3. Utilisation of the ligno-cellulosic biomass is probably still far away, but it represents the ultimate solution.

In each of these three steps the farming activity largely relies on the help of well-planned research.

To outline the first step problems, a comment will be given on the results of a recent survey which shows that the first phase will be essentially based on two crops: sugar beet and grains. This phase will be dominated by the necessity to spread all the year round the activity of the ethanol producing factories.

The research programmes in this phase must be mainly focussed on four points:

- Specialised sugar beet utilisation
- Prolonged storage of beets
- 'High-test molasses' technique
- Wet milling of dry grains

The second step could be based on two 'new' crops: sweet-sorghum and Jerusalem artichoke.

The research will help the sweet-sorghum culture to switch from fodder to cultivars with high yield in culms and also in grains, resistant to beating down and harvested in late fall, when the sugar beet season is over. Microbiology research must find new yeast strains, more suitable for the fermentation of the sorghum juice, while the agronomic research must help in the selection of the new varieties and in seed production.

Jerusalem artichoke is a very low demand culture. It fits very well lands unsuitable for sugar beets or for cereals. However, a serious problem must be solved by research: how to control the dormancy of the plant.

The third step, as stated, must be considered as the ultimate solution. Plenty of research programmes are already in progress. The success of ligno-cellulosic biomass utilisation depends completely on the results of research.

NEW PERSPECTIVES IN LARGE SCALE PROCUREMENT OF PULPS AND FIBRES
FROM EUROPEAN AGRICULTURAL PRODUCTS

F. BARNOUD and M. RINAUDO
University of Grenoble I,
Centre de Recherches sur les Macromolécules
Végétales, CNRS, BP 68, 38402-Saint Martin d'Hères Cedex, France

INTRODUCTION

As is well known, at the present time EEC members are largely dependent on North America and Scandinavian countries for the cellulose pulps used in the paper industry.

The wood pulps produced in the European Community represent a small percentage of consumption. In 1983 European production was only of the order of 5 000 000 tons of unbleached and bleached chemical pulps compared to imports of 20 000 000 tons. Straw pulp was only 300 000 tons although straw, as a residue of the cereals crops, is extremely abundant (5 to 8 million tons of straw remain without valuable use in France every year).

Hardwoods (several species coming from the so-called coppices) are also largely unutilised (according to the French forestry services there are 20 million cubic metres of hardwood in Europe).

This dependence of the EEC on foreign suppliers has two consequences: the high cost of pulp and the possibility of shortage of this resource in a period of international crisis. The continuous decline of European production of pulps over the last 20 years is due to several factors, the most significant being the non-competitive nature of the small units of production, most of which are old factories. The high degree of pollution caused in the atmosphere and streams by the conventional processes, kraft or sulfite processes, is also an unfavourable factor.

At the present time, the size of the non-European modern units of chemical pulps production is of the order of 1 000 to 2 000 tons per day instead of the value of 200-300 tons per day characteristic of the old European plants. A few large size units are functional in Europe. They produce pulps at competitive prices.

This situation of crisis is not unavoidable. New techniques of forestry, new species of wood and new methods of processing wood are now realities which could be greatly developed in Europe. A large number of medium size factories could transform the pulp industry in the EEC in the coming ten years.

Let us consider the new opportunity offered by sylviculture and pulp makers for the immediate future.

AGRICULTURAL WOOD FIBRES

Since 1970 remarkable researches have led to a completely new technique to grow wood. When growing poplars, eucalyptus, sycamores or Sequoia by the technique of 'short rotation coppices', a very high wood productivity can be achieved. In France the private company AFOCEL has shown recently by several reports, that this technique of growing wood can lead to an average production of wood biomass of up to 18 to 20 tons per hectare per year. Mechanisation and automation of the crop and the homogeneity of the product are positive factors for the pulp industry of the future. Such high productivity is obtainable on rich and moist soils. It must be considered

that this wood production is an alternative to cereals and mainly maize
crops. Unused lands can also be put to good use to produce this kind of
wood. Technical and economic studies are now in progress in our Institute.

THE INDUSTRIAL PROCESSES TO MAKE PULPS

The chemical treatments of the materials generally implies a
delignification step (kraft or bisulfite process) in which part of the
hemicelluloses are dissolved and the fibres separated and swollen. The pulps
for the paper industry are named chemical, semi-chemical or chimico
thermomechanical pulp (CTMP). A large family of high yield pulps can be
obtained now by using either conventional hardwoods (mixed species) or
Graminae stems like reeds (Arundo donax) or straw.

About ten years ago research was undertaken in the French 'Centre
Technique du Papier', in Grenoble, on the most appropriate method of
processing straw to produce pulps at a reasonable cost. This process is
characterised by the use of sodium carbonate as the delignifying agent.

The processes can be adapted for small units of 50-10 t/day without
severe problems of pollution. Hemicelluloses (xylans) are maintained in a
high proportion.

The use of corn stalks instead of straw stems to make pulp is the
source of severe technical difficulties due to silicium. These difficulties
have not been solved. Also the proportion of cell parenchyma is still higher
than that observed in straw. Accordingly, there is no possibility to use
such an important residue to make cellulose pulps.

In the present world situation of cellulose production there is a need
to make more investigations on new processes of delignification of
ligno-cellulosic residues coming from hardwoods or straw to produce either a
weakly delignified fibre product or a rather pure cellulose which can be
used in other industrial areas.

PRODUCTION OF NEW TYPES OF FIBRES OR OTHER ORGANIC MATERIALS

Natural ligno-cellulosic materials are normally exploited as a raw
material for the chemical industry: monomeric units are recovered after
total hydrolysis as monosaccharides or as phenols and these chemicals have
many industrial uses.

A second method of exploitation, which is less common and in need of
development, would be to use the undergraded polymers directly.

Polymeric materials from biomass can be used as fibres in the pulp and
paper industry, or transformed as filaments as in the viscose process. For
this purpose the cellulose is dissolved to a collodion extruded and
coagulated to produce a continuous filament of regenerated cellulose. It
must be noted that the viscose process is long, expensive and polluting by
comparison to the processes used in synthetic polymers.

Another possibility suggested by recent research concerns the use of
new direct solvents for cellulosic materials, such as amine oxides. New
adequate solvents to produce fibres without pollution have to be found in
the coming years.

Up to now cellulosic pulp of high grade (α-cellulose) is needed in the
viscose process to obtain a total dissolution. Taking into account the
composition of the different ligno-cellulosic materials it is clear that the
yield in α-cellulose and the ease of obtaining it depend on the morphology
of the fibrous material.

With new solvents it is hoped that the yield can be increased, for
example, if a given fraction of lignin is left. In our laboratory
ligno-cellulosic filaments can be obtained easily according to a new method
which will be patented.

It is also of interest that new products can be prepared from the biomass as a blend of synthetic and natural polymers; for such new polymeric systems mechanical properties have to be tested and compared to the initial viscose or other high value fibres such as Fortisan.

PRETREATMENT OF LIGNO-CELLULOSIC MATERIALS

To increase the solubility of ligno-cellulosic materials a pretreatment may be necessary. It will increase the accessibility to solvent. The steam cracking is described and some results allow to demonstrate that after a very fast treatment (around 1 minute) and a decompression phase (a pressure drop of about 40 bars) the hemicelluloses become water soluble and a large part of the lignin also becomes soluble in alkali or organic solvents.

The Stake process (Technip - France) is a continuous process for steam cracking. It is a rapid treatment which corresponds to small industrial units. It allows the separation of the main components of the biomass, cellulose, hemicelluloses and lignin. This process will favour:
-    the production of dissolving pulp for filament procurement,
-    the production of cellulose derivatives with different physical conformations (thread, film, foam, etc.) due to the increase of accessibility to reagents. We can also envisage the production of a new kind of fibre for the pulp and paper industry if the disruption of the fibre's structure can be avoided by a careful control of the decompression step.

More technical information about the destructuration of wood chips, the high yield pulps and the flash hydrolysis of wood can be obtained by inquiry to the Centre de Recherches sur les Macromolécules Végétales.

SEPARATION, EXTRACTION AND FRACTIONATION OF MILK PROTEIN COMPONENTS

J.L. MAUBOIS
Dairy Research Laboratory, INRA,
65, rue de Saint-Brieuc, 35042 Rennes Cedex, France

(Based on an article which first appeared in Le Lait (Nov.-Dec. 1984),
No. 645-646, pp. 485-495, and reprinted with permission)

Milk proteins, unique milk components belonging to the genetic patrimony of the various species, are essential nutrients in the human diet, especially for the newborn. Milk components already represent 20 to 30% of the total dietary proteins in the industrialised world (Hambraeus, 1982), which is an indication of the substantial use of milk and its derivatives in industrialised countries. It can be predicted that the usual type of milk consumption will increase slightly in the future (1 to 2% per year, according to the country) due to the differing price evolution of animal proteins - milk proteins are two to three times cheaper than egg or meat proteins, they have a better nutritional value and they are presented to the consumer in an infinite variety of forms, such as cheeses. It is clear that this price difference is being taken seriously by consumers in the context of the present world economic crisis. Consequently, consumption of conventional dairy products will increase, not only in the major dairy countries, but also in many other countries where governments are trying to raise the proportion of animal protein in the diet.

However, it is likely to be in the field of new products that potential openings for milk proteins will be the largest, the most diversified and the most able to lead to high valorisation. Indeed, thanks to the fantastic progress in knowledge achieved by dairy research, and to the emergence of extremely fine separation techniques which are very well adapted to the 'biological sensitivity' of these components, the dairy industry is now able, and will become more so in the near future, to produce a large variety of new protein products to serve the needs of downstream industries such as the food or pharmaceutical industries.

The purpose of this paper is to attempt to review all the new ways of transforming milk proteins. Such a review is not easy because the topicality of the subject has meant that access to some information was not readily available for obvious reasons of industrial property. Nevertheless, I have made the review as complete as possible.

Figure 1 is a schematic representation of all the new possibilities of separation, purification and fragmentation of both milk protein categories: caseins and whey proteins. It does not pretend to be complete, since it only concerns the major proteins, and obvious technological steps such as evaporation and drying are not indicated.

SEPARATION OF WHOLE MILK PROTEINS

Whole proteins can be isolated from other skim milk components (lactose, mineral salts, non-protein nitrogen) with two techniques based either on simultaneous precipitation of casein and whey proteins under the triple action of a high heat treatment (90°C - 1 to 20 minutes), reduction of pH (pH is brought down to 5.8; 5.3 or 4.6) and calcium chloride addition

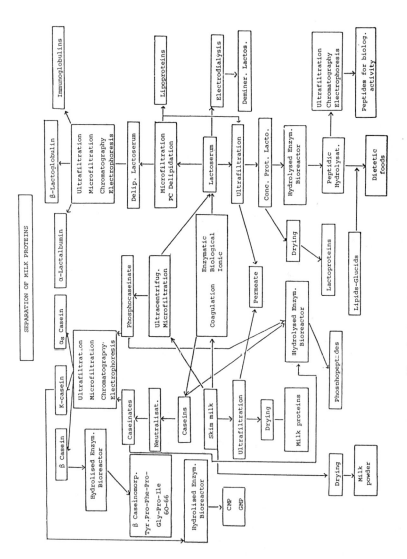

Fig. 1. Separation of Milk Proteins

(0.03 to 0.2%), or on selective retention of protein components by an ultrafiltration membrane.

The first process was initiated in the USSR, but subsequently developed in Australia (Muller, 1982). The resulting coprecipitates have a calcium content of around 0.5 to 3%, but their solubility is low. Membrane ultrafiltration (Figure 2) allows a large variety of milk protein-enriched products to be obtained with a protein content (Nx6,38/T.S.) ranging from 33% to 85%, a content of lactose and mineral salts that can be adjusted as desired by using diafiltration, and with a high solubility over a wide pH range.

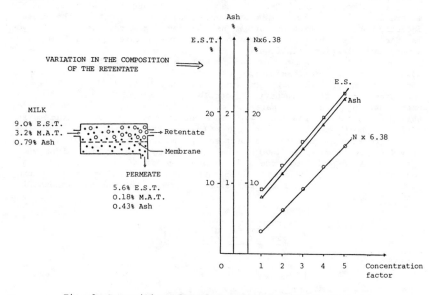

Fig. 2. Composition of Products Obtained Through Membrane
Ultrafiltration of Skim Milk

SEPARATION OF CASEINS AND WHEY PROTEINS

Past, present and future separation techniques of the two major groups of milk proteins can be classified according to the casein property used for extraction. This extraction technique can be physicochemical, biological or purely physical.

The oldest known technique for casein separation is the physicochemical one. It even supports the casein definition (Gordon and Kaplan, 1972). Lowering the milk pH to 4.6 leads to casein precipitation. The precipitate is washed several times in order to reach a satisfactory degree of purification. All types of acid can be employed as precipitants, but those most commonly used are hydrochloric and sulphuric acids. Because of poor valorisation of acid casein wheys, techniques of so-called ionic acidification (Triballat, 1979; Rialland and Barbier, 1980) have been developed recently. They are based on the exchange of milk cations (Na+, K+, Ca++) with protons (H+) using ion exchange resins. The resulting wheys have a lower mineral content, especially those resulting from the Bridel process (Rialland and Barbier, 1980) which do not contain any acid anions. Another advantage of the latter process is an increase of casein yield due to the retention in the curd of the main proteose-peptone as the consequence of an hysteresis effect of the solubility of this component (Pierre and Douin, 1984).

Casein separation through biological processes leads to products with very different properties and therefore with different ultimate valorisations. Lactic fermentation allows lactose bioconversion to lactic acid until pH is reduced to 4.6. The precipitate of casein obtained has similar properties to that of acid casein. The addition of rennet to skim milk allows the splitting of the K-casein fraction and so destabilises the casein micelles. Coagulation then takes place with the release in the whey of caseinomacropeptide (CMP). The resulting rennet casein is highly mineralised and has both plastic properties, of interest in the sausage industry, and stretching properties used in cheese-making. When formaldehyde is added before hot pressing, rennet casein precipitate is transformed into a very hard plastic: galalithe.

Separation of casein through physical techniques is still prospective. However, recent progress in porous materials, such as high mechanical resistance metallic alloys or composite materials, will lead to the emergence on the industrial scale of microfiltration and of ultracentrifugation for separating all the different types of casein from skim milk. Figure 3 is a schematic representation of a process for preparing native phosphocaseinate through the ultracentrifugation of an ultrafiltration retentate, as proposed by Maubois et al. (1974). Very promising results were obtained on a laboratory scale regarding yield (practically equal to the theoretical maximum), composition and concentration of sediment and supernatant, but the equipment manufacturing partner was unable to build the required continuous industrial ultracentrifuge.

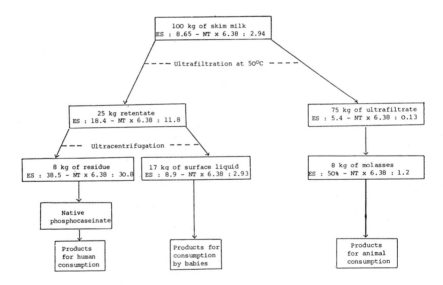

Fig. 3. Schematic Representation of the Process Combining Ultrafiltration and Ultracentrifugation for the Production of Native Phosphocaseinate (Maubois et al., 1974).

The recent commercialisation of mineral microfiltration membranes (Veyre, 1984) with pore homogeneity, which can be expected either because of the use of new ceramic materials (ZrO_2, AL_2O_3, Csi) or due to the use of high energy radiation for preparing thick screen membranes, leads us to expect that, in the near future, casein and whey proteins will be industrially separated by these physical techniques.

SEPARATION OF THE DIFFERENT CASEIN SPECIES

The objectives of such a separation are not based on caseins themselves - α_s casein 46%; β casein 34%; K-casein 13% - but on the products they allow to be obtained more easily and with higher purity. The casein on which most of the efforts of separation are devoted is β casein. Indeed, it could be used as the raw material for preparing β-casomorphin which is a heptapeptide located in the 60-66 position in the sequence. This β-casomorphin is similar to opiates or is a mediator for their synthesis, and could thus fulfil a primary role in sleep or hunger regulation, and in insulin secretion (Mendy, 1984). This explains the interest in this component for dietetic therapy. Industrial separation of β casein could be achieved in the near future with techniques such as microfiltration, ion exchange chromatography or continuous electrophoresis. Fragmentation in an enzymatic membrane reactor could be easy, but isolation of β casomorphin will require the use of new separation techniques, probably chromatography or electrophoresis.

It might prove of interest to isolate other casein fragments, either for their nutritional or even physiological properties, or for their functional properties. Recently, Shimizu et al. (1984) indicated that the N-terminal extremity (1-23) of α_{s1} casein had very good emulsifying qualities and could be easily separated from a peptic hydrolysate through centrifugation.

SEPARATION OF ORIGINAL SEQUENCES OF CASEIN

The presence of phosphoserine residues gives milk caseins a very marked chelating power versus calcium ions and trace elements. Physicochemical characteristics of native calcium phosphocaseinates lead us to consider that phosphopeptidic residues play an essential role in casein micelle stability, in mechanisms governing gel formation during coagulation and also in intestinal absorption of minerals and trace elements. These considerations led Brulé et al. (1980) to devise two patented processes for purifying these original casein sequences (Figure 4). Isolation of these products from peptidic hydrolysates is based on their chelating properties. Indeed, when calcium and phosphate ions are present in the solution, these peptides aggregate and thus can be purified with membrane ultrafiltration techniques, non-phosphorylated peptides going through the UF membrane. The sequestrating power of the resulting products is very high: 100 g can fix 5.6 of calcium, 10 g of copper, 5 g of zinc, 12 g of Fe++ or 5 g of Fe+++. This means that very large and diversified markets could emerge for phosphopeptidic products, and it is possible that European milk production will be insufficient to meet the demand in five or ten years time.

SEPARATION OF THE WHOLE WHEY PROTEINS

Whey is a diluted fluid containing 4 to 6 g of true proteins per litre. These proteins have excellent functional properties and a very high nutritional value due to their exceptional content in sulphur amino-acids, in lysine and in tryptophane. Extraction of these proteins for the purpose of human nutrition is not new, it is already done in the production of old whey cheeses such as Serac or Bruccio. However, it was only at the beginning of the 70s that, with the development of membrane ultrafiltration, a truly

new whey industry was born for preparing the very diversified whey protein
products required by downstream food industries (Maubois, 1982).

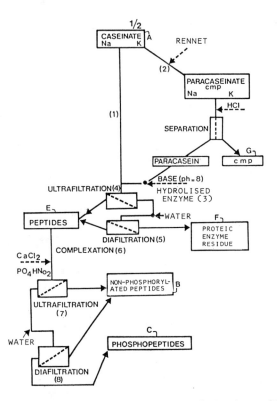

Fig. 4. Schematic Representation of the Preparation of Phosphopeptides from
     Caseinate (Brulé et al., 1980)

     Membrane ultrafiltration offers the possibility of preparing a large
range of whey protein concentrates (WPC) with a protein content from 35% to
85%. The main functional advantages of WPC are:

- solubility throughout the pH scale,
- high water retention capacity,
- gelification ability,
- foaming ability.

     Cheese wheys mainly, but also casein wheys to a lesser degree, contain
residual non-centrifugeable lipids which are responsible for opalescence in
the liquid. Most of these residual lipids are phospholipoproteins
(sphingomyeline, phosphatidylcholine and phosphatidylethanolamine) coming
from the fat globule membranes. Concentrated at the same rate as proteins
during ultrafiltration, their presence in WPC can limit market openings for
some applications, such as those requiring foaming functionality. They also

limit ultrafiltration fluxes and the efficiency of downstream fractionations and fragmentations. These lipoproteins could be specifically separated on an industrial scale in the near future, either by using the microfiltration technique, as proposed by Piot et al. (1984), or by using the physicochemical processes such as the one recently developed by Fauquant et al. (1985) which is based on aggregation of lipoproteins during a moderate heat treatment in the presence of calcium ions. Valorisation of lipoproteins thus extracted will be easy due to their excellent emulsifying capacities.

Removal of whey residual lipids is, in our view, strictly necessary before any attempt to separate whey proteins through chromatography. It is probably because this vital step was not observed, that the industrial scale-up of the Spherosil RP process has met its known difficulties. Indeed, whey lipoproteins have very marked amphoteric and amphiphillic characteristics, which lead to a strong adsorption on all porous materials. Consequently, this adsorption leads to fouling or even 'poisoning' of ion exchange resins, which becomes irremediable due to the physicochemical limits of cleaning acceptable by these materials. The application of anion exchange chromatography to defatted whey could allow, initially, secure preparation of WPC with a protein/T.S. ratio near 90-95% (Malige, 1982). Then, by varying eluting conditions (for example, use of pH or ionic strength gradients), the different whey proteins could be obtained.

SEPARATION OF THE DIFFERENT WHEY PROTEINS

In addition to the above-mentioned ion exchange chromatography, one can foresee the possibility of separating the individual whey proteins through techniques based on their differences in electrical charge by, for example, electrophoresis, or differences in molecular size by ultrafiltration or microfiltration. The latter techniques have led to very promising results regarding the separation of immunoglobulins and α-lactalbumin enriched fractions (Roger and Maubois, 1981).

The purpose of separation of immunoglobulins is in applications for protecting the intestinal tract of the young calf; the purpose of separation of α-lactalbumin is in the exceptionally high tryptophane content of this protein (4 residues per mol). Such a protein could allow preparation of tryptophane-containing peptides which could be used as precursors of serotonine (a neuropeptide which regulates sleep and hunger (Mendy et al., 1981)).

FRAGMENTATION OF WHEY PROTEINS

Because of their high nutritional value, whey proteins can constitute all the proteinic part in dietetic therapy. However, formulation of the product requires that the nutrients administered via the enteral route to patients suffering from major intestinal diseases have analogous structures to those of metabolites arriving in the human intestine after normal digestion. This means that limited operations of digestion which cannot be accomplished by the patient have to be achieved outside the human organism. Such a simulation of proteinic digestion can be carried out by using enzymic membrane reactor technology (Maubois and Brulé, 1982). The products leaving the reactor contain peptidic sequences able to allow not only an optimum absorption (Figure 5) (Rérat et al., 1984), but also to start, directly from intestinal receptors, operations of anticipating regulation (hormonal and enzymic secretions - guidance of metabolic crossroads). A regeneration or a compensatory hypertrophy of missing intestinal segments can be produced thanks to this type of feeding (Mendy, 1984). Fragmentation of milk proteins, studied by the Sopharga Society and our INRA laboratory in collaboration, is now a commercial reality (Reabilan is the trade name of the product), and the production will be scaled up to a level of several hundred tons next year.

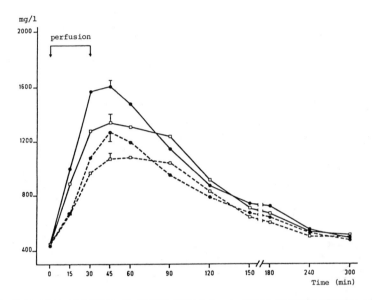

Fig. 5. Evolution of α amine in Portal Blood (——) and in Carotid Blood (---) after Perfusion of 55 g of Whey Protein Enzymatic Hydrolysate (●) and of 55 g of Amino Acids (□) (Rérat et al., 1984)

CONCLUSIONS - PERSPECTIVES
     The already available separation techniques, and those which will become available in the near future, offer dairy technologists fantastic tools for separating, fractionating and fragmenting the most interesting milk components for human feeding, i.e. proteins. It is in this field of 'proteinic cracking' that the dairy industry will take an important part of the 200 billion dollars market forecast for biotechnologies at the beginning of the next century. It has the ability, due to the profound knowledge of its raw material - milk, due to the advanced technology of its equipment, and due to the skills of its professionals. However, the dairy industry must be constantly aware of the need to progress through continuous innovations, not only in the specific dairy market, but principally in the global context of the glucidic, lipidic and proteinic components market, and in close collaboration with downstream industrial sectors.

REFERENCES

(1) BRULE, G., ROGER, L., FAUQUANT, J. and PIOT, M. (1980). Procédé de traitement d'une matière à base de caséine contenant des phosphocaséinates de cations monovalents et leurs dérivés. Produits obtenus et applications. Brevet français no.80 022 81.

(2)   FAUQUANT, J., VIECO, E. and BRULE, G. (1985). Clarification physico-chimique des lactosérums de fromagerie. Le Lait (submitted for publication).

(3)   GORDON, W.G. and KALAN, E.B. (1974). Proteins of milk. Fundamentals in Dairy Chemistry, 2nd Ed. B.H. Webb, A.H. Johnson and J.A. Alford (Eds), Avi Publ. Co., Westport, p.87-124.

(4)   HAMBRAEUS, L. (1982). Nutritional aspects of milk proteins, in Developments in Dairy Chemistry. 1. Proteins. P.F. Fox (Ed), Appl. Sc. Publ., London, p.289.

(5)   MALIGE, B. (1982). Les protéines de lactosérum extraites par chromatographie. Protéines animales. C.M. Bourgeois and P. Le Roux (Eds). Techn. Docum. Lavoisier, Paris, p.191-201.

(6)   MAUBOIS, J.L., FAUQUANT, J. and BRULE, G. (1974). Procédé de traitement de matières contenant des protéines telles que le lait. Brevet français no.74 39 311.

(7)   MAUBOIS, J.L. (1982). Les protéines de lactosérum extraites par ultrafiltration. Protéines animales. C.M. Bourgeois and P. Le Roux (Eds). Techn. Docum. Lavoisier, Paris, p.172-190.

(8)   MAUBOIS, J.L. and BRULE, G. (1982). Utilisation des techniques à membrane pour la séparation, la purification et la fragmentation des protéines laitières. Le Lait, 62, 484-510.

(9)   MENDY, F., BRACHFOGEL, N. and SPEILMANN, D. (1981). Actualités dans le domaine de la connaissance, de l'utilisation digestive et métabolique en nutrition humaine des protéines laitières. Rev. Lait. Franç., 400, 37-58.

(10)   MENDY, F. (1984). Fragmentation des protéines laitières. Biofutur, 24, 60-61. Interview de J. Rajnchapel-Messai

(11)   MULLER, L.L. (1982). Manufacture of casein, caseinates and coprecipitates, in Developments in Dairy Chemistry. 1. Proteins. P.F. Fox (Ed), Appl. Sc. Publ., London, p.315.

(12)   PIERRE, A. and DOUIN, M. (1984). Eléments d'étude du procédé Bridel de fabrication de caséine à partir de lait décationisé par échanges d'ions (E.I.), Le Lait, 64, 521-536.

(13)   PIOT, M., MAUBOIS, J.L., SCHAEGIS, P., VEYRE, R. and LUCCIONI, M. (1984). Microfiltration en flux tangentiel des lactosérums de fromagerie. Le Lait, 64, 102-120.

(14)   RERAT, A., LACROIX, M., SIMOES-MUNES, C., VAUGELADE, P. and VAISSADE P. (1984). Absorption intestinale comparée d'un mélange d'hydrolysats ménagés de protéines laitières et d'un mélange d'acides aminés libres de même composition chez le porc éveillé. Bull. Acad. Nat. Méd., 168, 385-391.

(15)   RIALLAND J.P. and BARBIER, J.P. (1980). Procédé de traitement du lait sur une résine échangeuse de cations en vue de la fabrication de la caséine et du lactosérum. Brevet français no.2 480 568.

(16)   ROGER, L. and MAUBOIS, J.L. (1981). Actualités dans le domaine des technologies à membrane pour la préparation et la séparation des protéines laitières. Rev. Lait. Franç., 400, 67-75.

(17)   SHIMIZU, M., LEE, S.W., KAMINOGAWA, S. and YAMAUCHI, K. (1984). Emulsifying properties of an N-terminal peptide obtained from the peptic hydrolyzate of α_{s1}-casein. J. Food Sc., 49, 1117.

(18)   TRIBALLAT, (1979). Procédé et installation pour la préparation de la caséine à partir du lait et produits ainsi obtenus. Brevet français no.2 428 626.

(19)   VEYRE, R. (1984). Utilisation des membranes minérales Carbosep en industrie agro-alimentaire. Le Lait, 64, 261-275.

OLEOCHEMICAL RAW MATERIALS AND NEW OILSEED CROPS

F. HIRSINGER

Henkel KGaA, Postfach 1100, D-4000 Düsseldorf 1,
Federal Republic of Germany

Summary

Fats and oils are important raw materials for oleochemistry as
well as for nutrition. Of the three main groups of natural oils and
fats, tallow serves with 70% as the most important raw material for
oleochemistry, followed by lauric oils (15%) and other vegetable oils
(8%).

The majority of these oils are not being produced in Europe. The
oleochemical industry, however, has a need for certain chemical
characteristics in its raw materials that can hardly be obtained from
European agriculture in its present state. The special chemical
requirements are: sources of high-concentrate oleic acid, or high
erucic acid and high lauric acid, oils with functional groups, and
waxes in seeds. Examples are given for alternative new oil crops as
well as 'old' crops that have been custom-tailored by plant breeding
for oleochemical needs.

The oleochemical industry is very supportive of such new methods
in European agriculture and would welcome more cooperation. This would
not only help with diversification in crop rotation but also help
prevent surplus production and subsidy.

Fats and oils are important raw materials for nutrition as well as for
the oleochemical industry. Twenty per cent of the world supply of oils and
fats (57 million tons) is being used by the industry. In Europe this amounts
to 1.7 million tons or 3% of the total world production; a comparatively
small amount (Figure 1).

Most of those oils, however, are not being produced in Europe. Even
tallow, almost the cheapest source of fat used by the industry, comes
preferably from non-European countries, because such imported tallow is of
superior quality. A similar situation is given for vegetable oils: **most** of
those oils like coconut oil, palm kernel oil and soyabean oil are produced
outside of Europe; of the fats and oils that are produced in Europe, only
rapeseed oil has some importance as a raw material for European
oleochemistry.

These oils are used as starters for quite a number of different
chemical pathways that lead to a whole bunch of highly diversified
substances. Some of the final products can actually be synthesised either
with petrochemical or with oleochemical raw materials. The oleochemical way,
however, is building on a generic base: almost unlimited natural resources,
whereas petrochemistry is based on limited fossil resources. When comparing
the chemical formula of paraffin and fatty acid, there appears a striking
similarity that explains their exchangeability (Figure 2).

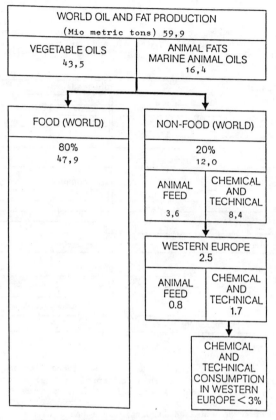

Fig. 1. Oil and Fat Consumption in 1982
Estimated in Million Tons

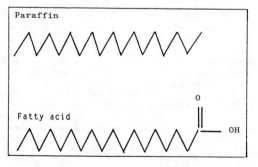

Fig. 2. Comparison of Chemical Formula

88

For understanding the importance of different oils it is necessary to explain their chemical characteristics. There are three main groups of oils: the first and most important group is that of long chain vegetable oils such as those of soyabean, sunflower, peanut and rapeseed. These oils consist mainly of fatty acids with a carbon chain length of C18 (Figure 3).

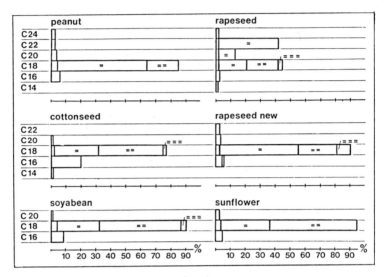

Fig. 3.

The main difference of the oil crops mentioned lies in the degree of saturation of their seed-oils. This is expressed by the content of oleic (18:1), linoleic (18:2) and linolenic acid (18:3). Unsaturated oils are very important for nutritional uses with special emphasis on linoleic acid, because the higher desaturated linolenic acid has a tendency to turn rancid quickly. The needs of the oleochemical industry for vegetable oils, however, are quite opposite. A high content of oleic acid is preferred because this is most suitable for certain consecutive chemical reactions. None of the oils mentioned, however, can supply this sufficiently. Normally oleic acid contents are not higher than 50%.

In Europe a certain exemption was given about ten years ago when rapeseed oil still contained up to 40% erucic acid. Nowadays erucic acid still has its importance in chemistry but needs to be imported from other countries instead.

The second group of oils is that of short and medium chain fatty acids from vegetable oils such as those of coconut and palm kernel (Figure 4). These oils can only be produced in tropical countries. They consist mainly of lauric and myristic acid. They have high prices. Some of their physical properties are also desirable for certain confectionery products. Their oleochemical use is most important as raw material for detergents, soaps, emulsifiers, etc. With 46% they are the most important raw material of oils for classical oleochemical uses.

The third group of fats is that of animal tallow and fish oils. Most important of those is tallow which consists mainly of palmitic, stearic and oleic acid with oleic acid contents around 40% (Figure 5).

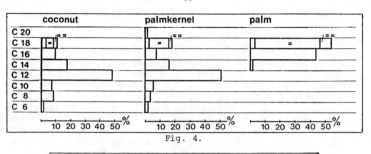

Fig. 4.

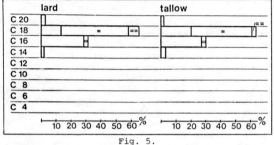

Fig. 5.

As mentioned before, such a low percentage of oleic acid is not what the industry prefers. There are, however, two advantages which contribute to the outstanding importance of tallow as an oleochemical raw material: its comparatively low price and, most importantly, an easy physical method of separating oleic from stearic acid: the hydrophylisation process, which enables oleic acid contents of more than 70% (Stein, 1968).

The overall fatty acid pattern of all the natural oils and fats would look as indicated in the upper part of Figure 6.

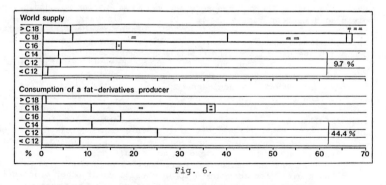

Fig. 6.

Such an imaginary 'world oil' would consist mainly of long chain fatty acids of the first group mentioned previously. Its medium chain fatty acid percentage would be less than 10%. Today's consumption of a large fat derivatives producer, however, is 44% medium chain fatty acids.

What are the main characteristics of the oleochemical raw materials? Often they are by-products that are no longer useful for nutrition, such as tallow, castor oil or oils that contain erucic acid. Out of these, high value products are being synthesised. In addition, oils of high quality such as coconut, palm kernel and soyabean oil are needed.

In most cases these oils do not have optimum fatty acid patterns, no matter whether they are by-products or of high value. The reason for this is the fact that those plants have never gone through intensive breeding work for oleochemical adaptation. Because of this, these oils have to be cleaned and separated with a lot of technical input. This means that there will always be a necessity to find by-products or other inexpensive sources of oils and fats. Moreover, the segregation of the different fatty acids plus the glycerol itself, result in new by-products that have to be marketed as well. Altogether these processes and products have to be competitive with those fatty derivatives that can be synthesised from petrochemical raw materials.

Most important for this particular meeting is to remember that only very low amounts of European grown oils are used by European oleochemistry. The main reason for this is not only their chemical constitution but also their comparatively high market price. With this I do not want to criticise EC price policy. I just want to emphasise that there must be a shift to improved oil qualities, if those prices are to be justified. I am even convinced that there might be considerable opportunities for European agriculture, if it can satisfy special oleochemical needs. Chemistry needs fatty acids of certain chain length, whereas for nutritional purposes different oils and chain lengths can be substituted. This is actually one of the main differences between the food and the chemical industry: while the one is mostly interested in physical properties of the raw material, such as colour, taste and smell, the other is solely interested in its chemical composition (Richtler and Knaut, 1984).

Up to now all the efforts in producing the raw material - oil - have been directed towards nutrition. European plant breeders and agronomists should be aware, however, that oleochemical needs might be something interesting to look into. There are actually signs on the horizon which indicate that overproduction of conventional oils and fats does not necessarily need to happen in the EC. To give some examples:

- In the USA some years ago new varieties of sunflowers were released by the Sigco Corp. These lines were derived from Russian mutants. Their oleic acid content ranges from 85 to 92%. This year these varieties have been grown over an area of 20 000 hectares in the US (J.R. Smith, private communication, O.I.L., SF). Without many problems such lines could also be grown in southern European countries.

- A similar oleic acid content of 85% can be obtained from the seeds of Euphorbia lathyris (caper spurge), a plant that has a considerable history in Europe (Hondelmann and Radatz, 1983). After its introduction to mid-Europe by the Crusaders, it was used for centuries as burning oil for lamps. Early this year a German breeder successfully applied for variety protection through the federal seed inspection for the first variety of Euphorbia to be released. Its seed oil content is around 50%, the seed yield around 4 000 kg/ha, which means that 2 000 kg of oil/ha can be realised!

- Breeding stations in West Germany are looking at high erucic rapeseed varieties again, for adaptation to industrial uses. Other cruciferous crops like Crambe abyssinica, Indian mustard and Eruca sativa might yield even higher percentages of erucic acid. Such new developments deserve much more interest from agriculture when considering that higher prices can be obtained for such custom-tailored new sources of industrial raw material. Moreover, this would help diversify European agriculture.

-    A project that was launched ten years ago in the US in the state of Oregon was aimed in this very direction of crop diversification. The need for crop diversification was actively supported by the Oregon Department of Environmental Quality, which developed a policy against the common field burning technique in grass seed production by asking farmers to pay a fee for field burning. With this money new crops development was supported. Out of this the Meadowfoam project was initiated (Jolliff, 1981). This crop (Limnanthes alba) synthesises in its seeds a wax that can replace sperm whale oil. It is a winter annual crop that is by far less complicated to manage than Jojoba. In 1985 about 400 hectares of Meadowfoam were grown by Oregon farmers. The crop should be well adapted to middle and southern European climates too.

-    Another candidate for special industrial needs is Cuphea. This plant can synthesise in its seeds medium chain fatty acids with high lauric or very high capric acid contents; much higher than in any of today's commercial lauric oil sources (Figure 7).

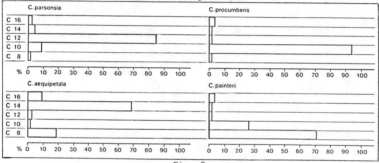

Fig. 7.

Domestication of this crop has never really been accomplished and will probably also take its time, because those wild species of plants of Mexican origin represent some characteristics that are not easy to overcome. Among these, seed shattering seems to be most disadvantageous. Field research under favourable climatic conditions, however, has shown that this crop is manageable.

A feasible harvesting technique was developed by simply sucking the seeds with a vacuum from the matured plants. This method can also be applied by mounting the blower of an air conditioner to a tractor power-take-off. Oil yields of 400 kg/ha were obtained with such simple harvesting equipments in California (Hirsinger, 1984, 1985). Certainly this crop of subtropical origin is not well adapted to middle European conditions. As with some of the other new crops mentioned, it should, however, be able to perform much better under the more favourable climatic conditions in southern Europe.

At the present time, the majority of these new oil crops have not yet been developed into new sources of industrial raw materials. They have been identified, however, as having high potential and they need to be drawn to the attention of plant breeding and agronomy. The oleochemical industry is very much interested in such new developments and is very supportive to them. More cooperative efforts with state departments as well as agricultural marketing organisations need to be initiated for successfully introducing those new oil crops to European agriculture.

REFERENCES

(1) HIRSINGER, F. (1984). Cuphea, die erste annuelle Ölpflanze für die
     Erzeugung von mittelkettigen Triglyceriden (MCT). Fette, Seifen,
     Anstrichmittel 82, 385-389.

(2) HIRSINGER, F. (1985). Agronomic potential and seed composition of
     Cuphea, an annual crop for lauric and capric seed oils. J. Am. Oil
     Chem. Soc. 62, 76-80.

(3) HONDELMANN, W. and RADATZ, W. (1983). Euphorbia lathyris L. as a
     potential crop plant. An outline. Angew. Botanik, 57, 349-362.

(4) JOLLIFF, G.D. (1981). Development and production of meadowfoam
     (Limnanthes alba). New Sources of Fats and Oils, 9, 269-285.

(5) RICHTLER, H.J. and KNAUT, J. (1984). Challenges to a mature industry:
     Marketing and economics of oleochemicals in Western Europe. J. Am. Oil
     Chem. Soc., 61, 160-175.

(6) STEIN, W. (1968). The hydrophilization process for the separation of
     fatty materials. J. Am. Oil. Chem. Soc., 45, 471.

ALTERNATIVE USE OF THE LAND FOR MEDITERRANEAN CROPS

V. BIANCARDI[*]

Istituto di Economica e Politica, Agraria della Facolta Agraria,
Via Filippa Re, 10, I-40126 Bologna, Italy

INTRODUCTION

In 1972 the Liquichimica Biosintesi factory in Calabria produced food from oil by employing modern biotechnology in an atmosphere of enthusiasm and optimism. Two years later the oil shortage occurred and the business was forced to close.

Today fuel is produced from food. The world has changed and will continue to change. There is a risk of solving problems which have already been solved perhaps because reliable, alternative energy sources will be found to replace oil, or viable substitutes to replace ethanol or lead.

CHANGES IN CONTEXT

Several innovations have recently caught the attention of the agricultural sector. With government officials worried by increasing surpluses, these new factors have given rise to hopes for radical changes which would enable agriculture to realise its full production potential without a sea of surpluses.

The first of these factors is a change in perception. Having reached, and even regularly exceeded, self-sufficiency in many foodstuffs, EEC agriculture can now look towards totally new, non-food products for use as raw materials in such industries as chemicals, pharmaceuticals, textiles, and building. As it did in food production, agriculture may well reveal much the same potential with regard to products for industrial use. It might even be said that only the imagination is capable of giving us an idea of just what and how many goods can be manufactured with the thousands upon thousands of very special molecules that nature provides.

The second factor is biotechnology research and the wide range of high-value, primary and secondary metabolites using media from various agricultural products, including non-food products, which can be derived from them. The innovation resides in the fact that, unlike former methods which could produce only directly edible plants, these techniques now enable many plant types to be grown which will provide raw materials for food or other useful raw materials, such as fuels, drugs, surface-active agents, rubber or fibres. In fact, it is the type of molecule, rather than the raw material from which it derives, that will be of paramount importance in future. Cereals, potatoes and beans will be replaced by plants grown for their starch, proteins, oils and fats, all of which will be chosen from a much wider and more varied range than that presently available.

Such a situation will open new markets and multiply the uses of agricultural production and enable it to sidestep the 'stomach barrier' of Engel's law. Moreover, as these products are renewable, they will help dispel the spectre of shortage of energy as well as of food and they will ensure the support of public opinion in this concept of agriculture. Besides these effects of product diversification and use, biotechnology methods will also induce large increases in yields.

----

[*] The author acknowledges the help and assistance of Marco Genghini and Michele Pontalti in the data programming and processing.

The problem, then, is one of abundance: whether diversification of use will prevail over yield increases. To date, and without biotechnology input, the FEOGA guarantee costs have continually increased (Table 1). Their incidence on the added value in agriculture has increased by 50% in eight years (1975-83), from 0.35 to 0.5, and today equals the yield from 12 million hectares of cultivated land, that is two-thirds of Italian farm land.

Table 1 FEOGA-Guarantee Cost Incidence on G.d.p.

Years	G.d.p. at market prices Billion ECU	FEOGA COST Million ECU	G.d.p. %
1975	1 107	3 903	0.35
1976	1 269	4 414	0.35
1977	1 415	4 692	0.33
1978	1 569	6 394	0.41
1979	1 763	8 297	0.47
1980	2 017	9 312	0.46
1981	2 210	9 394	0.43
1982	2 430	9 721	0.40
1983	2 690	13 434	0.50

* Net Cost

Source: EUROSTAT

The effect of biotechnological innovation will depend on the context in which it occurs. Advances in biotechnology will reduce costs, without affecting the trend in surpluses, only if an adequate price system exists which can respond to yield variations induced by these same innovations. This is the only way in which biotechnology will restore competitiveness and efficiency to EEC agriculture. Research, development and implementation of such innovations can bring about this restoration, but only if the will exists to do so. If not, a non-competitive agricultural food sector will be joined by another, equally non-competitive sector in the production of raw materials for bioindustries. It is unthinkable that raw materials at a price much higher than world rates can lead to competitive industrial products without subsidies. The opportunity afforded by biotechnology to return EEC agriculture to competitive levels will either be taken or lost; the problem of surpluses alleviated or aggravated.
These changes all occur within a framework of:
1.   Increasing competition in world markets due to:
     a) The US decision gradually to decrease agricultural prices, and
     b) The transformation of certain countries from traditional importers to net exporters of agricultural produce.
2.   A shift in EEC agricultural policy away from unlimited guarantees.

EEC AND MEDITERRANEAN SURPLUSES
The greatest share of the financial burden for the EEC's agricultural policy can be ascribed to continental products (70-75%), whereas, as Table 2 shows, the cost incidence concerning the Mediterranean products has grown at a greater rate.

Table 2 EEC: Cost Escalation % Per Sector

(FEOGA-GUARANTEE)

SECTOR	1974	1983
Cereals	12.4	15.6
Rice	0.3 (a)	0.5
Sugar	3.4	9.0
Olive oil	3.5	4.3
Seed-protein crops	0.5	6.8
Horticult. produce	1.9	6.8
Grape vine-wine	1.3	4.0
Tobacco	5.4	4.2
Other*	0.7	0.4
Milk and dairy	40.7	29.8
Meat**	22.0	13.4
Price Subs.-Curr.	16.2	2.6
TOTAL	100	100
VALUE (Million ECU)	3 094	15 861

* Seeds, hops, apiculture
** Beef, pork, sheep-goat, eggs, poultry
(a) 1973 datum

Source: EUROSTAT

The reasons are manifold. There has been a series of bumper grape crops. But the increase in production has not been spontaneous; it has resulted largely from improved management.

In the ten-member Community the Mediterranean surpluses which have been withdrawn from the market are in citrus fruit, tomato, cauliflower, small quantities of table grapes and other fruits and vegetables. The data are in Table 3.

Table 3 Horticultural Produce Withdrawn from Market
(Metric Tonnes and Total Yearly %)

PRODUCE	1979/80	%	1980/81	%	1981/82	%	1982/83	%
Apple	548 938	7.89	517 798	7.33	53 733	1.07	1 146 932	13.43
Pear	54 131	2.57	162 926	6.96	120 677	5.35	90 510	3.80
Peach	111 090	6.61	55 620	3.35	343 936	15.22	239 656	10.89
Orange	2 737	0.16	101 091	6.61	73 243	2.95	126 914	5.71
Mandarin, etc.	78 215	36.14	53 025	27.91	82 032	27.99	14 653	5.58
Lemon			21 755	3.05	70 253	7.08	160 386	19.76
Table grape			530	0.03				
Cauliflowers	40 732	2.94	13 217	0.91	12 069	0.73	40 108	2.36
Tomato	197 100	3.06	78 878	1.30	56 121	0.72	54 380	0.70
Eggplant							32	0.008
Apricot							343	0.10
Total	1 032 943	4.52	1 004 840	4.43	812 064	3.27	1 873 914	7.08

Note: 1979/80 9-member EEC
      1980/81 Include Greece for 6 months
      1981/82 10-member EEC
      1982/83 10-member EEC

There are also surpluses in such produce as wine and fruit (mainly peaches) which are not limited to the Mediterranean countries. For these products it is obviously difficult to detect whether the surplus areas are in the north or south, even though the superior quality of the wine, its yield disparities (1 hectare of vineyard on the continental lowlands produces as much as 5-30 hectares in the Mediterranean areas), as well as the early-ripening of many fruits and vegetables, should leave little doubt in identifying those areas characterised by greater yield capacity, thus excluding most of the Mediterranean countries from any responsibility for surpluses.

In addition, especially in the south, there are crops which, while not surplus, are subsidised and are thus a burden on Community finances. These are olive oil, durum wheat, oil-producing vegetables (sunflower) and protein-rich vegetables (broad beans, vetch, lupin).

THE SPECIFIC FEATURES OF MEDITERRANEAN AGRICULTURE

The specific features of the Mediterranean areas which are significant in determining alternative land use are:

a)   Farm structure, and
b)   Environment.

a) Mediterranean agriculture is chiefly characterised by many very small farms (4 million farms smaller than 5 hectares) of intensive cultivation (Table 4). They produce mainly fruit and vegetables rather than livestock and employ a higher number of farm labourers (6 million) (Table 5). In addition income is very low as compared to that of continental areas (index = 64).

Table 4 Farm Number by Acreage (Mediterranean Countries)

Country	1980	
1 to 5 ha		
France	234 000	
Italy	1 501 000	*
Greece	519 000	*
Portugal	600 000	**
Spain	1 020 000	**
10-member EEC	2 645 000	
5 to 10 ha		
France	165 000	
Italy	377 000	*
Greece	151 000	*
Portugal	-	
Spain	459 000	(a)
10-member	974 000	
TOTAL		
France	1 135 000	
Italy	2 192 000	*
Greece	732 000	*
Portugal	800 000	**
Spain	1 700 000	**
10-member	5 670 000	

* 1977 datum   ** European Parliament session Report 1-785/82
(a) See note (a). Includes Farms up to 20 ha.
Source: EUROSTAT

Table 5 Agricultural Employment (Mediterranean Countries)

COUNTRY	1970	1980
Total Agr. Employ. (000)		
France	2 821	1 841
Italy	3 878	2 925
Greece	1 279	1 016
Spain	3 587	2 122
Portugal	997	1 120
12-member EEC	16 633	12 882
Agr. Employ. % of total		
France	13.5	8.5
Italy	19.6	13.9
Greece	38.8	29.2
Spain	29.5	18.9
Portugal	31.7	28.5
12-member EEC	11.2 (a)	11.0 (b)

(a) 10-member EEC
(b) 1977 datum
Source: EUROSTAT

Table 6 Farmer Income Level by Region

Member Nation	Region-Effect
Germany	77-131
France	79-220
Italy	31-132
The Netherlands	250
Belgium *	197
Luxembourg *	110
UK	84-208
Eire *	96
Denmark *	208
Greece	41-55
10-member EEC	31-250

Note: 100 = Community net value added mean for farm/agricultural work unit
* These nations form a single 'Region' for RICA

SOURCE: European Community Commission - Agriculture in the EEC - 1984 Report

On the one hand, it is this same abundance of labour which, together with the chronic patterns of land holdings and distribution, impedes both farm re-organisation and technological innovation (mainly in terms of cost reduction), and on the other hand, induces the farmers to pursue and extend intensive crops, thus conditioning the choice of an acceptable, alternative land use. These farms also pose the greatest problems in terms of the number of people involved, the employment policies related to them, and the obstacles confronting any modification of the property system. In addition, although there are also medium and large size farms, especially in Spain, for which crop innovation is perhaps easier and includes a broader range of alternatives, their contribution to the production of surpluses is minimal.

b) From the environmental point of view, the most outstanding features of the Mediterranean areas are probably the variety of microclimates, of soil conditions and of topology. While natural differences are at times attenuated and at others enhanced by man, on the whole the environment is characterised by:

1. Greater amount of sunlight.
2. Higher temperatures with occasional frosts and wind.
3. Greater evapo-transpiration.
4. Less rainfall (especially in summer).
5. Shorter growing season in arid zones under field crops and a longer one in irrigated areas either under orchard or particularly resistant field crops.
6. Decentralised position with respect to the large European markets.
7. Frequently atypical soils situated along sparse coastal lowlands and in vast sub-arid, mountainous inland areas.

Generally speaking it follows that there is: a greater potential of photosynthesis; markedly reduced limits to, and risks inherent in cold-susceptible crops, and their increase for drought-susceptible crops; the need of less energy input for protected crops, with consequent cost reductions; a greater need of water input in the irrigated areas or, if unavailable, the replacement of irrigated crops by hardier, more resistant species; greater transport costs and difficulties and limitations for the more perishable, fresh-market crops.

The present agricultural landscape of the Mediterranean areas is the result of the response over many generations and the combination of these different factors. It is a layout delimited by flexible boundaries and greater risks in comparison to the continental farming map, especially where water for irrigation is lacking. It is therefore a landscape more closely bound to traditional crops.

The typical crops may be grouped as follows:

a) Grains (cereals, seed-oil and protein-rich crops)

They are viable as climate-imposed indigenous species (e.g. durum wheat and sunflower) and when grown on large, intensive-cultivation farms.

Difficulties arise when they must compete with continental crops and when grown on small family-run farms. Formerly grown for self-consumption, they are now being rapidly replaced (400 thousand hectares less in 10 years in southern Italy) by intensively cultivated crops as these farms become more market-oriented.

b) Fresh-market fruit and vegetables

Not only are these crops well suited to the Mediterranean environment (early and late ripening and lower energy costs for protected crops), but to the farms themselves, which are geared to labour-intensive cultivation and a prolonged work cycle. Such crops are therefore better able to respond to the expectations of a surplus work force eager to increase its income. However, they are perishable, expensive and difficult to transport, and there are problems linked to their short marketing period and limited international appeal.

c) Processed fruit and vegetables (wine, oil, tomato)

These are high quality products which are not normally affected by competition from continental crops. Wine and oil are also striking and unparalleled examples of intensive land use in sub-arid conditions. Unfortunately their market is restricted although it might be possible to extend their outlets.

d) Livestock

Less important than crops, livestock is being cut back everywhere except in some mountain areas. Whereas cattle breeders are generally less competitive and are yielding to fruits and vegetables, both sheep and buffalo breeding have better prospects. This is because their products are typically Mediterranean (main as well as by-products) and in demand, thus creating a favourable market situation. Moreover, unlike cattle, they require only low amounts of forage crops (soya and feed in general) which are scarce in Europe.

Generally speaking, the agribusiness farmers, and especially those operating in fertile areas, have proved remarkably dynamic in adopting the most suitable innovations, particularly early ripening vegetables and most protected crops, including those requiring a high degree of specialisation. In many instances these new crops seem to have found an ideal environment, enabling them to be grown at markedly lower costs and harvested at increasingly earlier dates as a result of intensive management methods, varietal breeding advances and technical innovations. In contrast, the farmers on poor soils have encountered greater difficulties, obviously due to the lack of water, lack of alternatives, isolation, soil characteristics and acreage extension and mechanisation problems. However, there are instances of the non-use of valuable resources which are difficult to explain, especially as to irrigation water.

The salient characteristics of Mediterranean agriculture are small farm size, the trend to mixed crops, the widespread distribution of grape vines and olives, and scarce irrigation. In southern Italy seed crops cover an area equal to 46% of the utilised agricultural area, whereas permanent and forage crops equal 54%. There is also a marked decline in cattle breeding in central (-34%) and southern Italy (especially in the lowlands where it is practically non-existent), which contrasts to that in the mountain areas, where it registered a significant increase (+22%), and on the islands (51%). Obviously livestock is not a valid alternative to grape vines and olives, since it thrives where the latter do not.

ALTERNATIVE LAND USE OF MEDITERRANEAN LAND

At least hypothetically, many alternatives exist. They range from non-use (land bank) and non-agricultural use (parks and recreation) to the development or introduction of non-food and the spreading of non-surplus food crops, especially those most suitable to the environment.

All solutions, however, pose problems. Non-use is a political choice made possible only by a socio-economic context which is not normally found in Mediterranean zones. Better prospects are offered by the creation of parks and recreational areas. This is due to both the great demand for parks in the vicinity of every medium and large city and the ensuing creation of jobs, such as attendants, for the ex-farmers. However, new parks would have little effect on the reduction of surpluses.

Then, too, there are those food crops in which the EEC is not self-sufficient and which could be expanded. Among such crops are corn (+20-25%), some minor cereals (+30-35%), rice, some fresh fruits, lamb and mutton (+30-35%), horse meat, nuts and especially several seed-oil and protein crops. Any decision to expand the acreage of these crops (and especially the last) must, however, take into account the self sufficiency ratios (Table 7) and the gap which already exists between EEC and world market prices and the resulting international political implications.

The Mediterranean environment provides favourable conditions for the growing of some protein-rich crops (broad beans, lupins, vetch). They have all the advantages of legumes (cheap fertiliser) and do not require special implements. They do, however, cause several problems which are mainly linked to the irregular cropping, moderate yields (due to an as yet unsatisfactory breeding and inadequate knowledge of pests), long rotation, and the uncertainty of long term price levels artificially supported by large Community subsidies.

Table 7 Production, Imports, Exports and Self-sufficiency Ratio for the Main Farm Products Within the 10-Member EEC (1982-1983). Data in Hundred Metric Tonnes Unless Otherwise Stated

Product	Production	Import	Export	Reserves	Consumption	Ratio
Common wheat	55 619	2 253	13 566	3 652	40 654	136.81
Durum wheat	4 049	1 210	1 414	55	3 790	106.83
Barley	41 101	146	5 863	1 295	34 089	120.57
Rye	2 412	56	2 300	- 101	2 546	94.74
Maize	19 728	5 567	905	229	24 161	81.65
Oats and other	7 686	140	108	49	7 669	100.22
Other cereal	420	131	6	25	570	73.68
TOTAL CEREAL	131 015	9 503	21 885	5 154	113 479	115.45
Rice	724	613	292	9	1 056	68.56
Sugar (1)	13 492	1 493	3 091	445	9 474	147.16
Potato	33 960	394	1 021	259	33 074	102.68
Wine (2)	171 935	5 079	9 006	12 871	155 137	110.83
Apple	7 162	569	254	161	7 316	97.90
Pear	2 089	120	113	12	2 084	100.24
Peach	2 169	12	120	0	2 061	105.24
Fresh fruit	18 080	3 948	1 178	20	20 831	86.79
Citrus fruit (3)	3 516	5 561	755	0	8 322	42.25
Cauliflower	1 469	4	5	0	1 467	100.14
Fresh tomato	7 735	439	91	0	8 085	95.67
TOTAL VEGETABLES	32 697	2 965	2 947	0	32 898	99.39
Milk (4)	28 658	12	134	0	28 536	100.43
Butter (4)	2 091	108	400	306	1 640	127.50
Beef (4)	6 663	374	392	20	6 616	100.71
Pork (4)	10 183	111	226	9	10 097	100.85
Sheep-goat meat	706	281	4	- 26	975	72.41
Eggs	4 264	33	162	5	4 128	103.29

Source: EUROSTAT, Agriculture in the EEC, 1984 Report

Note: (1) Total production
(2) Production in thousand HL
(3) Including jams and juices
(4) 1982 data

Therefore, although conditions are theoretically good, in practice these crops are declining on the most precarious farms, at least in southern Italy. These are, in fact, small farms which need to raise their per hectare output. Thus, these basically extensive crops are not effective replacements for existing ones, except on large and intensive-cultivation farms in non-surplus areas.

In the less fertile zones sheep breeding should be encouraged. Its characteristics and products (milk for cheese rather than wool) in the Mediterranean clearly distinguish it from that of continental Europe.

Hopes of effective replacements can be based on non-food agricultural crops which, together with the food crops, EEC and Italy especially, import at a rate which significantly aggravates the trade deficit.

The share of the non-food deficit is important in terms of magnitude (6 thousand billion Lire as against 9 thousand billion Lire for food products) for it involves raw materials used by the country's most important industries. Some experts believe that this deficit can be at least partly reduced by placing a part of the surplus acreage under non-surplus crops. Others suggest that the same result might be achieved by improving the use of marginal areas, consequently helping to keep people on the land.

Other non-food sectors where Italy is even further from self-sufficiency are timber, leather, suede, animal and certain vegetable fibres, natural elastomers, special vegetable oils (e.g. linseed oil), paint oils, etc.

Great hopes are also being placed in the new non-food crops as well as on the different uses of traditional crops. In the latter instance, stimulating prospects have been offered by the EEC decision progressively to decrease the lead content in petrol and the intention of eliminating the gap between EEC and world market prices of starch for the chemical industry.

The high cost of cultivating these industrial-use crops could be overcome through priority subsidies granted to farmers of the most costly surplus crops already subject to production quotas, such as milk, in order to promote replacement. The areas designated for the new cultivation could be identified on the basis of surplus crop acreage distribution and substitute crop efficiency rating. They should be concentrated, that is, so as to allow the installation of efficient processing plants, which are directly related to a maximum reduction of such costs as transport, servicing, consulting fees.

It should, however, be pointed out that the crops used for alcohol production as well as the other non-surplus food crops (corn, soya, etc.) are generally more suited to continental rather than Mediterranean environments. It is, therefore, likely that they will become more effective substitutes for continental (milk, sugar) rather than Mediterranean surpluses, as their introduction in the latter area would create problems affecting soil fertility, farm incomes, employment and farm stability. Although the effect of such replacement crops on Mediterranean surpluses would be indirect, the ensuing result would in no way be diminished (e.g. the replacement of vineyards in the north would also solve wine market problems in the south).

The discussion surrounding new crops suited to the Mediterranean environment has recently focused on several species. Of these the jojoba has sparked the keenest interest as an excellent substitute for sperm-whale oil. Although its development in Italy is still confined to a few score experimental hectares, the country possesses a fairly large extension of acreage climatically suited to this crop. This acreage alone could provide yields exceeding the foreseeable EEC market share. It has indeed been estimated that potential world acreage is in the order of several hundred thousand hectares.

This means that even should criteria such as crop adaptation and profitability be met, its influence on alternative land use in the Mediterranean areas of the EEC would be far from resolving the problem.

Another new crop in this regard is the guaiule, an herbaceous plant which provides a product equivalent to natural India-rubber. According to our estimates, 150 thousand hectares of this crop would be sufficient to meet all of Italy's rubber requirements.

Other crops proposed as substitutes are cochia, used as animal feed but of dubious value with respect to traditional products, and fejoa, a fruit similar to the kiwi which would make it competitive with other fruits. Cyperus, an herbaceous crop which can be mechanically harvested, yields small tubers rich in protein and an oil very similar even in taste to olive oil. Kenaf, a fast-growing herbaceous plant, produces fibre for the newsprint paper industry. Mention should also be made of sorghum and, especially, the Jerusalem artichoke (for making sweeteners or as a raw material in the low-cost production of ethanol). Although there has been some support in Italy for the Jerusalem artichoke as a crop suited to marginal areas, both it and sorghum are primarily efficient fertile soil crops.

For all of these new, industry-designated non-food crops, adequate information on cultivation techniques and yields in Mediterranean areas is still insufficient. It is consequently apparent that surveys, research, agronomical experiments, technological studies and economic evaluations will have to be undertaken to ensure their viability and feasibility.

The prospects offered by medicinal plants and herbs are far less concrete than is commonly believed, despite dramatic consumer market expansion in Italy and the EEC.

There are several East-European countries (especially Hungary) with particularly favourable climatic conditions (many wild plants, skilled gatherers, etc.) which have achieved a high degree of specialisation and a solid market share. This makes it all the more imperative to evaluate accurately the environmental, technological and economic factors upon which any market competition with these traditional producers will depend.

Further, it should be considered whether, as some suggest, these crops are more suited to marginal areas or, as row seems probable, they are only competitive and profitable when intensively grown in fertile soils with modern technology. One should also take into account the possible reactions of 'ecologist' consumers vis-a-vis a no longer wild but intensively cultivated product, chemically treated and hence not natural.

Any attempt at solving the problem of Mediterranean surpluses must also include possible expansion of non-deficit crops (flowers and plants) which are particularly suitable to this environment. The advantages are particularly apparent in the case of protected and greenhouse crops, since they allow considerable energy savings (e.g. up to 2 kg of diesel fuel per kg of tomatoes).

The cultivation of these crops should be promoted and supported. Suitable structures and infrastructures must be created, research improved and the results made available.

Secondly, it should be possible to exploit more fully certain special and quality crops such as olive oil, wine, and other typical products. The 12-member EEC will in fact be the exclusive producer of oil and wine (76% and 60% respectively). It is hard to believe that the rest of the world (93% of world population) is too small a market to absorb our surplus wine (15% over domestic consumption), olive oil with price supports promoting both domestic consumption and exports, and other traditional Mediterranean foodstuffs at remunerative prices.

Market difficulties can be overcome if the quality of these products is taken into account. They are among the finest quality processed foodstuffs (e.g. cheese, salami, etc.) and, as in the case of extra fine olive oil, have a high nutritional value.

Before taking steps to replace these and other products, new and more aggressive marketing strategies should be studied. They are excellent products largely unfamiliar in world markets and hence still without competitors, not to mention their dietary importance as pointed out by important medical associations (e.g. COMA, Committee on Medical Aspects of Food Policy). Reconsideration of their market situation within the EEC might also be timely and appropriate. As it is, many northern countries subject wine to restrictions (excise duties), leaving it open to illegal practices and unfair competition from artificially enriched products (sugaring). It and other products are also not adequately protected by legislation (e.g. misleading labels on various types of olive oil and the absence of origin and quality denominations).

CONCLUSIONS

The alternatives then are many and varied. While no single one can solve the problem, taken together they can do a great deal to change the present situation.

Many of the new crops are alternatives to current ones, yet it is difficult to predict the extent to which the family-run farms will respond to them. That this response is important can be seen by the fact that these new crops are less labour intensive and could conceivably provoke an even more serious kind of surplus, i.e. labour, thus creating unemployment. For it must be remembered that, particularly in the south, alternative land use is aimed both at eliminating surpluses as well as increasing farmers' income, yet without creating unemployment at least in the short term. This can be achieved only by replacing present crops with ones having higher economic returns and similar labour-input levels. This means too that any long term solutions must focus on creating employment and land mobility opportunities as the conditions sine qua non in implementing the technical and crop modifications best suited to the area.

The decision makers should therefore base priorities and initiatives on introducing and improving what may be defined as the 'deserving innovation' and the 'valuable existing', respectively. Research is the basic and indispensable key to such a policy but must be undertaken immediately, as its low-cost and far-reaching results require time.

Since the main innovations reside in the growing and use of new and existing crops for industry, every effort should be made to programme supply and demand. This will enable industry to let farmers know what they need so that the farmers will know what to grow. Planning of this type would be greatly facilitated by establishing one or more scientific research coordinating or liaison groups between the two sectors. This could then be followed by more intensified plant-breeding and technico-agronomic research programmes capable of quickly transferring discoveries from the laboratory to the field, selecting the most suitable acreage and structures for cultivation, and evaluating the economic viability of development. Although such a system will undoubtedly require a great deal of time and energy, the success or failure of any replacement project will depend on its implementation.

In the final analysis the Mediterranean is a world which, far more so than others, calls for specifically designed measures to achieve effective and long lasting goals. For it must be remembered that this is an agriculture imbued with millenia of tradition, culture and civilisation.

AN HOLISTIC APPROACH TO EVALUATE THE POTENTIAL PRODUCTIVITY OF
UNCONVENTIONAL CROPS

C. EERKENS
Centre for Agrobiological Research
Bornsesteeg 65, PO Box 14, 6700 AA Wageningen, The Netherlands

Summary

Emphasis is put in this presentation on raw materials for durable products, that is products physically stable to be handled and traded on the world market, while allowing stock to buffer for temporary surplus and timing of sales.
Research on crops should not only increase the cultivation efficiency of conventional crops but should also include domestication of crops that match the potentials of up-to-date know-how in the food processing sciences and balance energy benefit appraisals of crop produced chemicals with chemicals synthesised in the realm of petrochemical industries.
While an holistic approach is suggested, for setting the sights of future crop sciences, the type of research would stimulate the challenge for entrepreneurs to develop new agri-processing industries.
Genetic efficiency of crop species and their varieties will be discussed, desiring more knowledge on formation of chemicals - energy content, value and variations under stress application - photo (morphogenetic) efficiency, desirable ratios for root, shoot, product storage organs, as well as (fossil) energy input/output efficiency of crops. It is contended that ways and means can be found to produce more valuable crop constituents when the performance characteristics of still undomesticated crops are exploited.
Although systematic work still has to be started, even on many conventional crops, a few unconventional crops are mentioned that seem to have potential value when viewed holistically. There should be an EC-strategy for an inter-institutional databank to register the genetic expressions - crop production characteristics - performed in various standardised or known climates and standardised rhizo-conditions.

1. INTRODUCTION
    The participants of the EC-Workshop 'Old and New industrial crops - their processing and feasibility'[1] clearly demonstrated that the productivity of crops is more than a quantification of dry matter. The complexity of factors that ultimately determine the value of a product would therefore require an holistic approach of analyses. Consequently, the efficiency of production will have to include efficiency of land use, averaged over many seasons, keeping control over healthy land, as well as efficiency of crop production exploiting the climate characteristics, cultivation technology and intrinsic quality of the material - i.e. valued by processors and consumers. In other words how much capital would be generated by a crop in pre- and post-harvest processes.

Some global balances of energy investments in various crops and efficiency of synthesis are discussed in Chapter 2, while some promising new crops are evaluated in Chapter 3. Suggestions for incentive programmes to support new entrepreneurs and establishment of a crop performance databank are mentioned in Chapter 4.

This paper intends to emphasise those products that can be stored long enough for easy trading and handling, also facilitating 'Europe's third world charity programmes' to purchase durable food to keep poor people free from hunger.

'New crops' in this paper include renewed old crops, such as, for instance, double-zero rapeseed - i.e. crops with a new property as a result of breeding or manipulation.

2. THE EFFICIENCY OF ENERGY INVESTMENTS OF CROPS

2.1. General Scope

For the illustration of an holistic analytical approach, determining the value of a new crop, it is inevitable that some elements receive more emphasis than others. The crops' own energy husbandry for maximal production is emphasised here, disregarding for instance control of stress factors such as pests, diseases, weeds and drought. The intention of this chapter, thus, is to press for research activities that require strengthening and more attention in order to make an holistic approach possible, rather than carrying a complete review of all important elements of current (ongoing) research.

Yet also, the boundaries of fossil energy inputs - fuel for cultivation machines, fertilisers and processing - would actually require a more careful assessment in this regard, eventually to be established by politics[2], in order properly to evaluate the energy output. In other words, the question should be answered whether it is permissible in the long term to maintain an energy input-output ratio of primary production in Europe higher than 1.0. (At present it is 1.25 in the Netherlands.)

Another aspect neglected in this paper relates to the appreciation of sensoric values of raw materials such as colour, grades, smell, taste - aspects, however, that will be automatically encountered in the realm of the processing economy at the factory.

2.2 Production Losses Caused by Environmental Factors

The average yield of most of the conventional crops is about three to four times smaller than record (maximum) yield as Boyer[3] indicated. In Table 1 the percentages of losses are subdivided by the contribution of diseases, insects, weeds and environmental losses. Herewith a clear indication is obtained about the order of magnitude of losses, pointing to the fields of research that need further exploration.

Table 1 Record Yields, Average Yields and Losses after (3)
         (in Parenthesis Estimated Dry Matter)

	Record yield ton/ha (RY) (DM)		Average yield % of RY	Losses by diseases insects	Weed losses	Losses land use cultivation + climate
Corn (maize)	19.3	(16)	24	7	3	66
Wheat	14.5	(12)	13	3	2	82
Soyabean	7.4	(7)	22	5	4	69
Sorghum	20.1	(17)	14	3	2	81
Potatoes	94.1	(21)	30	14	1	54
Sugarbeets	121.0	(22)	35	11	3	51

It has been stated by the author[4] that the estimated losses in <u>oil crops</u> caused by cultivation in sub-optimal climate conditions - thus excluding inappropriate land use, drought and poor farming technologies - could reach as much as 40%. It is speculated therefore that other commodity crops have also been promoted in areas where they cannot maximise their superior performance characteristics. Furthermore, when taking into account that 'breeders' can do very little to the improvement of photosynthesis mechanisms within the species[5] - as was evident for wheat, corn, sorghum, millet, sugar cane, cotton and cowpeas - a strong argument is created to search for other species, for instance those that have built up a natural evolutionary efficiency, possibly to be found in the realm of (persistent) weed populations.

2.3 Energy Husbandry in the Plant System

The levels of energy collected by photosynthesis of a plant and the subsequent usage, for various processes in the organs, are of crucial importance for the evaluation of productive cropping. Since it is impossible, so far, to determine directly the amount of joules taken, for building a specific plant organ, an attempt is made here to relate energy husbandry with amounts of carbon matter formed in various plant parts. In doing so, the collection of energy - roughly related to kg CO_2 uptake per hectare - can be balanced against the weight and value of carbon matter stored in the organs at harvest time.

Although the differences in efficiency for collection of solar radiation energy would also depend on the morphogenetic design of the leaves in the green canopy[6], for the purpose of comparing crops here the assimilation capacities of the leaf surface (LAI) are taken as a measurement for it. In Figure 1 an example is given of a few crops that photosynthesise at their maximum in different temperature regimes (FAO).

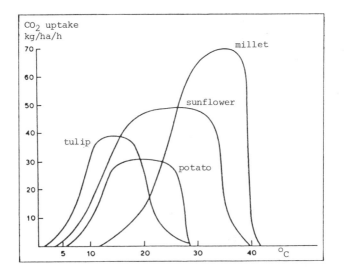

Fig. 1. Photosynthetic Productivity at Light Saturation on the Leaf Surface of Various Crops in Relation to Temperature

For a correct evaluation of efficient use of 'energy' it is also important to know how much of it is used for the synthesis of products c.q. respiration for various processes and maintenance of the plant structure[7].

Particularly respiration under high temperature conditions would consume a large part of the collected energy, so that the evaluation of efficient use requires a more complex approach. Among others Sibma[8] visualised the 'energy' distribution for a grass crop in average Dutch conditions, as given in Figure 2. Although the respiratory losses during hot days may reach a level of 55% of the total CO_2 uptake, the average losses during the whole life of the crop is estimated to be 25% of the total collected amount, leaving for 'energy' investments in leaves, dead material and roots respectively 35, 20 and 20%.

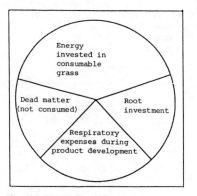

Fig. 2. 'Energy' Expenses by a Grass Crop,
Averaged Over Growing Season

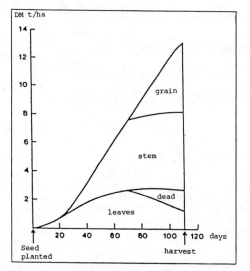

Fig. 3. Dry Matter Development Over Time of a Rice Crop

   With a rice model, as an example, it can be illustrated (Figure 3) how a plant distributes dry matter over time in various organs. This model, however, should not be considered as a fixed growth pattern, because Van Keulen and De Wit[9] indicate that with a low temperature regime during the juvenile stage of growth in rice, the invested energy proportions can be different and demonstrate that there are meteorological regimes, whereby the same amount of heat units, accumulated during growth, could yield more DM than in the average season - e.g. April versus December plantings in Bangladesh.

   Although the dry matter distribution of assimilates is not correct, for expressing energy investments, for a general approach of energy accumulation it is practical to illustrate which ultimate dry matter distribution is realised for a range of crops. In this regard Figure 4 is self-explanatory. When reviewing the value of a crop in an holistic approach, these types of analyses would be very helpful for a 'first order' efficiency screening, whereby the strategy relates to selecting crops that invest little energy in 'useless' organs.

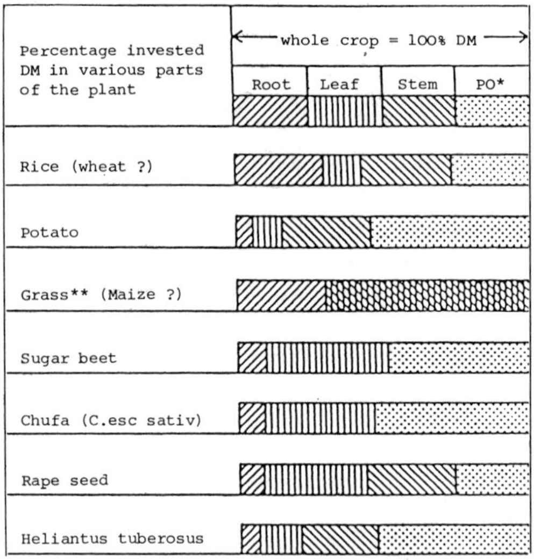

* PO = Principal organ (tuber, grain)
** PO in grass is leaf + stem (herbage)

Fig. 4. Approximation of Dry Matter Distribution in Plant Organs

2.4 Production Efficiency of Crops

   The level of dry matter produced in the various plant parts so far informs us about a global energy distribution strategy of the plant. There is a considerable lack of knowledge, however, about the plants' energy cost made for the synthesis of the final product as well as the determination of the value of the energy level of that product. Subsequently in this latter problem there are two elements discernible that require research attention:
1.   What proportion of the respiration energy expenses is utilised for the production of specific proteins, lipids, lignins, organic acids and carbohydrates, and

2. What value do these components possess for the agri-industries, viewed from the energy benefits in processing?

When reviewing the first problem it was observed by Eerkens[4] that the non-photo processes in senescing plants can lead to considerable improvement of the raw materials.   It was suggested that once the enzymes are triggered (signalled) to produce,   for instance,   oleics or linoleic lipids,   a higher temperature level is required for the transformation of erucic acid into the higher valued oleic acid in rapeseed oil,   as is exemplified in Figure 5. Consequently it was argued as to whether these processes can be induced in the harvested organs (seeds, tubers) or whether they need the osmotic thrust of assimilates in the senescing plant. The knowledge of the specific amounts of energy utilised by the plant for this 'upgrading' of chemical content should help to sort out process efficiency in the plant.

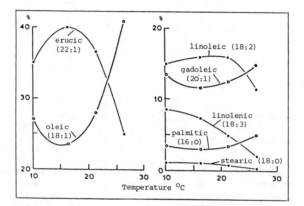

Fig. 5. Effect of Temperature During Growth on
Fatty Acid Composition of Rapeseed Oil

In order to appraise the second element, e.g. the value of the raw materials for (bio)industrial processing, it is desirable to determine the intrinsic energy level of the plant product.   In this regard Penning de Vries [10a] suggests an approach of determining production values of crops in accordance with the amount of glucose (primary energy source) used to synthesise the various groups of chemical constituents in the biomass - pointing out a method whereby these values are related to energy levels in carbohydrates, proteins, lipids, lignins and organic acids. While Penning de Vries (1984) calculated production values as being 1 211 carbohydrates + 1 700 proteins + 3 030 lipids + 2 119 lignins + 0 906 organic acids it is suggested by Vertregt [10b] to correlate production values with the total carbon content.   Accordingly a highly significant regression equation is obtained, e.g.: production values linearly related to carbon content.

In Table 2 the C-content of a number of interesting crop products has been summarised and, for easy reference, the chemical composition of intrinsic constituents tabulated so as to make one and another comparable. It should be noted that the data on the compositions are averages, approximated from scarce literature data, in other words just indicating an order to magnitude, and should not be taken as absolute figures.   It is contended that a refinement of this approach, in further research, could indicate subtle - yet very valuable - energy differences within the lipids, proteins, carbohydrates and lignins and subsequently be useful for prospecting in industrial processing economies. In this respect lost energy, for the production of undesirable constituents in a crop, will require also energy investments for their removal during product processing at the factory - for instance, extracting glucosinolates from rapeseed proteins!

Table 2 Percentage of Chemical Constituents of Raw Materials and Carbon Content

Species	Carbo-hydrates	Protein	Lipid	Lignin	Org. acids	Minerals	Carbon content (Cc)
Faba bean	55	29	1	7	4	4	49.2
Ground nut	14	27	39	14	3	3	63.6
Corn (maize)	75	8	4	11	1	1	49.2
Potato tuber	78	9	0	3	5	5	45.7
Rice	76	8	2	12	1	1	48.6
Wheat	76	12	2	6	2	2	47.6
Sunflower	25	14	42	13	3	3	56.1
Chufa (Cyp. esc s.)	53	6	30	6	3	2	65.2
Soyabean	29	37	18	6	5	5	55.8
Lupin seed	30	40	16	8	4	3	56.1
Hel. tuberoses	87	4	1	3	2	3	45.9
Rapeseed	30	23	38	6	3	5	63.0
Tulip	84	6	2	5	2	1	46.4

Yet, in conclusion, it should also be possible to express production efficiency (Pey) of the crop, with the above defined values, by introducing a link to a seasonal (climatological) reference - a factor reflecting the seasonal (daily) value of the land (Sdv). Although this issue is still on trial and of a 'brainstorming' nature (debatable on using instead, effective temperature, heat units and/or joules radiation) it is suggested here to take reference of Sdv as being 100 times the sum of (daily) potential evapotranspiration (Ep) of a green grassfield divided by the total annual Ep. For the middle of the Netherlands Ep can be roughly taken at 0.3 mm for a day in January, 0.5 mm for February, 1.0 for March, 1.8 for April, 3.1 for May, 3.5 for June, 3.3 for July, 3.1 for August, 2.2 for September, 1.3 for October, 0.7 for November and 0.3 for December. Consequently, when annual Dp = 640, has a Sdv of 100, then January gives a

Sdv of $\frac{31 \times 0.3}{640} \times 100 = 1.5$ and June: $\frac{30 \times 3.5}{640} \times 100 = 16.4$

A tulip season (November 1 - July 1) would have a Sdv = 53 and a potato season from April 15 - October 15 would have a Sdv = 80. When C is the carbon production in 1 000 kilograms per hectare

$$Pey = \frac{C}{Sdv} \cdot 100$$

Subsequently, when the maximum dry matter production of potatoes is taken at 21 tons per hectare and of tulip bulbs at 18 tons, then the Pey of potatoes will be

$\frac{21 \times 45.7}{80} = 12$, while the tulip Pey $= \frac{18 \times 46.4}{53} = 16$

Likewise, based on maximum yield estimates, one would calculate a Pey of 9 and 11.5 for wheat, respectively 00 (double-zero) winter rape. One would thus conclude that the tulip crop is more efficient in the Dutch climate than wheat, rape or potato crops, if this method were to be duly acceptable. In this regard it would be interesting to reflect on Pey values of a whole range of crop products. So far, the information gathered on short notice unfortunately lacks either sufficient Ep data or information on yield - e.g. not indicating maximum, average, actual or optimum - so that at this moment it is impossible to produce them.

3. A SELECTION OF SOME PROMISING NEW CROPS

3.1 General

From thousands of food producing crop species only a few have been domesticated to satisfy the world's demand. For instance, 80% of the world production of vegetable oils (40 million tons) is covered by five oil crops - soyabeans, palms, sunflowers, rapeseed and groundnuts - while we know that hundreds of other oil producing species exist that may have superior characteristics once domesticated for cultivation and production; it should be understood that the sketched approach in Chapter 2 is an attempt to begin with activities for a systematic inventory of crop characteristics. The method has not matured yet and requires further development, better data collection systems and a more precise definition of the value of various groups or specific chemical constituents. Therefore, the screening and selection of a few promising crops here below should be seen as an example - the best choice from a very incomplete list - rather than a contention that the selected crops are the best candidates for development on this latitude.

Yet no considerations are made for the prognoses of the 'net added value' to the farm of the future. In this respect commercial farming might need an overhaul of systems and technologies, so as to include for instance 'processing at the farmer's level' for semi-finished products, as was, among others, suggested by Rexen and Munck[12]. In turn, the industrial process sciences should be attuned in regard of these developments. In other words new technologies should be developed requesting inventive entrepreneurial activities.

Although the potentials of new fodder crops and tree crops (for example, production of maple syrups) should be included in this evaluation, they have been omitted so far.

3.2 New Bulb and Tuber Crops

Apart from above mentioned tulips, to be grown for their bulbs, two possible future tuber crops are suggested here, i.e. Chufa and Jerusalem artichoke. Chufa was already known by Egyptian Pharaohs and is of the family of the feared weed Cyperus esculentus, still cultivated on a small scale in Spain for its chufa milk (a health beverage). Jerusalem artichoke, Helianthus tuberoses, is one of the remainders of American-Indian agriculture.

With an estimated Sdv of 60, Chufa's production efficiency is calculated to have

a Pey of $\frac{65 \times 14}{60} \approx 15$

(DM production is based on extrapolation of maximal yield over a two year experimental period, i.e. = 13.8 tons!). Since the assimilates of the leaves are directly transferred to the tubers - e.g. stems or other storage organs are not made - and the vigorous growth behaviour in the family of Cyperus weeds is observed to be excellent, the perspectives for large scale development are likely to be good. In this regard Leemans[13] estimated that effective breeding could further improve the species to a yield of 16 tons per hectare within 10 years. Subsequently this could bring Chufa to a production efficiency (Pey) of 17. Moreover, when reviewing the perspectives for industrial processing Bevers[11] suggests that, for instance, 'pot extraction' developments, to collect the oil and sugars, could be relatively easily realised.

With respect to Jerusalem artichokes, some early estimates indicate that the production efficiency would approach the level of potatoes or even be somewhat higher. In a recent symposium[14] many of the processing qualities of the product have been discussed resulting in strong recommendations to develop this species into an industrial crop. The production of inuline, particularly, seems to have a high value for processing industries and may eventually become a potential competitor of high fructose corn syrups (HFCS). Apart from composites like Jerusalem artichoke and dahlia, other plant families like Liliaceae (tulip, hyacinth) and Amaryllidaceae (narcissus) are endowed with the production of polyfructosides and should be screened for their potential production efficiency.

In regarding the tulip bulbs as a potential starch producer very little is known about the perspectives for industrial processes. However, although the maximum production could reach 18 tons DM per hectare[15] there is some evidence for believing that bulb weight increase, by breeding, may drive production to a level of 20 tons per hectare, in which case production efficiency Pey would be more than 16.

3.3 New Seed Crops

In the realm of grains a lot of improvements have been sorted out over the past decades in relation to wheat, maize, barley, millet and soyabean. Yet, it is suggested that considerable improvements can be obtained in efficient land use by renewed introduction of improved species such as: peas, faba beans, double-zero rapeseed[17], lupins, falseflax, poppy, false ragweed (Iva xanthiifolia), lathyrus[16], evening primrose and cuphea (Cuphea lanceolata).

Only a few notes on some of these crops are reflected here. According to Röbbelen and Hirsinger (Universität Göttingen), cuphea is an excellent oil crop for producing butter-like vegetable oils, containing 80% C 8:0/ C 10:0 lipids in the oil fraction. So far no reliable productivity figures are known of it but, in accordance with some rough estimates of experiments, Pey figures of 14 to 15 can be reached which, on top of the premium value of the oil quality - hopefully in the future to be calculated by a refined methodology of energy level differences within the chemical group of lipids, possibly to be provided by vegetable oil processors - may make this crop a serious candidate for large scale development.

New lupin varieties without alkaloids are likely to produce the same sort of oils and proteins as the soyabean, while production efficiency is much higher than for soyabean in the colder climates. It is contended that this crop might compete in the future with double-zero rapeseed - another good candidate for development - particularly since lupins would not require nitrogen fertilisation.

4. CONCLUSIONS AND RECOMMENDATIONS

Among others, with the help of a rough scanning of the literature, it is concluded that there is a shortage of knowledge about the efficiency in plants of producing chemical constituents in relation to environmental factors. Systematic data collections on production characteristics of plant species are virtually non-existent. When data collections do exist there is generally a shortage of objective references, i.e. with respect to specific conditions of growth and/or crop variety registers. Most investigations at research stations relate to supporting further improvements of domesticated

crops while evaluation of intrinsic values of wild species - their prospects
and perspectives - are neglected because potential benefits are not reviewed
as a whole, i.e. efficiency for industrial processing, efficiency for
prospective farming technology and efficiency in genetic properties (the
right crop in the right climate).

Many large food industrialists do consolidate efficient - patented -
process technologies that are geared to (a few) well known crops, while
considering the developments of new raw materials as a threat to their
established investments. Consequently, the logical tendency to the monopoly
of a few sorts of crops by industrial giants, could seriously hamper the
promotion of new crops and thereby jeopardise the future innovative vigour
of Europe's food industrial development. It is recommended therefore that EC
entrepreneurs, submitting a sound holistic investment plan for the
exploitation of new crops (for instance Chufa), should obtain ample support
to get their new industries, with new raw materials, established.

On the basis of recommendations by the author[4] it is suggested to
establish a centre for the collection of basic data of crop performance.
Crop production characteristics and production efficiency data could be
obtained from different climates in standardised greenhouse procedures - for
instance, in the way that Shell Research is considering the establishment of
a greenhouse complex in England - or alternatively from systematic annual
collection of data from crops grown in standardised substrate solution in
the open air, for instance, hydroponics in sand. The dual purpose of such a
central databank is to provide a means for entrepreneurs to assess new crop
production efficiency as well as for bio-industrial researchers to obtain
understanding of (enzyme) reactions of processes in the plant in their
relation to stress, periodic changes of temperature, humidity, light
(climate).

One of the details of the holistic approach relates to establishing a
method for measuring crop production efficiency in various climates. It is
recommended to proceed with an approach as indicated in Chapter 2.4, and to
test the suggested methodology for its usefulness.

Finally there is evidence for believing that the EC policies for
subsidies and protection of farm products need to be reconsidered since -
apart from enhancing surpluses - they could hamper the above mentioned
developments of new crops and new industries.

REFERENCES

(1)   Proceedings EC Workshop 'Old and new industrial crops - their
      processing and feasibility'. (1985). Wageningen, IBVL, 8-10 November.
(2)   BAKKER, Th.M. (1985). Eten van eigen bodem. Een model studie.
      Dissertation to obtain Ph.D., LEI-institute.
(3)   BOYER, J.S. (1982). Plant Productivity and Environment, Science
      Vol.218, 29 October.
(4)   EERKENS, C. (1983). More efficient use of genetic properties in oil
      crops by exploitation of climate characteristics. Proceedings (1).
(5)   KAHN, M.A. and TSUNODA, S. (1970). Evolutionary trends in leaf photo-
      synthesis and related leaf characteristics among cultivated wheat
      species and its wild relatives. Japan. J. Breed. 20 (3), 133-140.
(6)   GOUDRIAAN, J., VAN LAAR, H.H., VAN KEULEN, H. and LOUWERSE, W. (1984).
      Simulation of the effect of increased atmospheric CO_2 on assimilation
      and transpiration of a closed crop canopy. Zeitschrift der Humboldt-
      Universität zu Berlin, Math.-Nat. R. XXXIII.
(7)   PENNING DE VRIES, F.W.T. (1975). The cost of maintenance processes in
      plant cells, Ann. Bot. 39, 77-92.

(8)   SIBMA, L. (1985). Private communication, Centre for Agrobiological Research, Wageningen.

(9)   VAN KEULEN, H. and DE WIT, C.T. To what extent can agricultural production be expanded- CABO Pub. N3 367, Wageningen, Options IAMZ - 84/1.

(10a) PENNING DE VRIES, F.W.T. (1975). Use of assimilates in higher plants. Cambridge University Press.

(10b) VERTREGT, N. and PENNING DE VRIES, F.W.T. (1985). A rapid method for the determination of the energy content in the biomass of plants. (Bepaling van de energie-inhoud van plantaardige biomassa). Publication in preparation.

(11)   BEVERS, J. (1985). Cyperus esculentus: Onkruidkundig probleem of technologisch fenomeen. Agricultural University, Dept. Food Sciences, Wageningen, Referaat 25 March.

(12)   REXEN, F. and MUNCK, L. (1983) Total utilization of crops in a balanced production of food. Feed, and industrial products - a vital task for biotechnology in the EEC. Carlsberg Research Lab. Copenhagen, Proceedings (1).

(13)   LEEMANS, J.A.L. (1984) Retired botany specialist of Unilever Research Vlaardingen, Private communication.

(14)   FUCHS, A. (1985). Inuline, een uitdaging;
       VERVELDE, C.J. Aardpeer als plant en als gewas; teeltveredeling;
       WUSTMAN, R. Bewaring van aardperen, IBVL: Themadag Inuline, June.

(15)   BENSCHOP, M. (1980). Photosynthesis and respiration of Tulipa sp. cultivar Apeldoorn, Scientia Hortic. 12: 361-375.

(16)   HONDELMAN, W. and DAMBROTH, M. (1983). Evaluation of seedoil producing wild species. Pflanzenbau und Pflanzenzüchtung der Bundesforschungsanstalt für Landwirtschaft, Braunschweig - Völkenrode, Proceedings (1).

(17)   BOURDON, D., PEREZ, J.M. and BAUDET, J.J. (1984). New types of rapeseed oil meals with low glucosinolate content. Station de Recherche sur l'Elevage des Porcs, INRA, St. Gilles, F-35590 - L'Hermitage, France. EC Workshop on rapeseed, Copenhagen, September.

DISCUSSION PANEL 1 : CHEMISTRY OF GLUCIDS

Chairman: W.F. RAYMOND (UK)

1.1 Introduction

The Panel noted the comprehensive review of this subject already published by the Commission (Cereal Crops for Industrial Use in Europe; Rexen, F. and Munck, L.; EUR 9617EN, 1984, 242 pp). It was thus able to concentrate its discussion on identifying important subjects on which current research information appears to be inadequate.

1.2 Sugars

Sugars, in particular glucose, are already used in many industrial processes; the potential use in ethanol production is evaluated by Panel 2. However their utility as an industrial raw material can be limited because polysaccharide hydrolysates generally contain mixtures of sugar monomers. The Panel considered that research is needed on improved methods for separating and purifying the monomers from hydrolysates and also on the potential offered by the saccharides as building blocks for the synthesis, especially of novel polymers (chirality). Sugars are also highly reactive, and more precise methods of selective reaction are needed in order to increase yields of specified intermediate chemicals obtained from sugars without protection of functional groups like hydroxymethylfurfal, dianhydrohexitols and the chemistry of their derivatives.

The utilisation of some sugars can also be limited because important micro-organisms lack the specific enzymes needed to ferment them; thus Saccharomyces cerevisiae, which has potential for producing important nutritional and pharmacological products, is unable to ferment lactose, currently a surplus product of only limited value. The transfer to S. cerevisiae of the missing enzyme system (β - galactoside permease) from for example Kluyveromyces lactis, is an example of research which could widen the range of uses for surplus sugars.

1.3 Starch

About 2/3 of the current starch production of 3.5 million tonnes in the EEC comes from maize, most of which is imported. While some of this could be substituted by Community-grown maize, the main potential appears to be in wheat starch.

However if wheat is to make a bigger contribution to starch production more information is needed on the composition and quality of the starches produced from different wheat varieties grown under a range of environmental conditions, and their suitability for different industrial processes. In particular information is needed on the relative values of the amylose and amylopectin fractions in starch, and the extent to which their proportions differ in different wheats, as an indication of the need, and the scope, for breeding for improved starch quality in wheat (as has already been done with high-amylose maize: Rexen and Munck, p.70). Similar attention should be given to the starches which can be extracted from barley, pulses, potatoes and Jerusalem artichoke. The Panel recommended that the Commission should organise a specialist seminar on starch, aimed at identifying and defining future research requirements.

1.4 Polysaccharides

The Panel noted the important developments that have recently been made in rapid methods for the hydrolysis of polysaccharides, which strengthen the case for improved methods of monomer separation, noted in 1.1. They considered however that the main potential use of polysaccharides, particularly those of a fibrous nature, will continue to be in the paper and board industries. Equally they recognised that, although there is considerable scope for replacing imported fibres, in particular of wood pulp, this change is unlikely to be initiated by the industries themselves, many of which are closely integrated with the overseas suppliers of fibre; any significant increase in usage of 'home-grown' fibre will require official stimulus. However before such replacement is considered it must be shown that it would be economically sound.

Thus the Panel concluded that a priority requirement is for a critical, independent, evaluation of the limited number of straw-pulping plants already operating in the Community, e.g. at Frederica in Denmark, and of the several new pulping processes now under development (Rexen and Munck, p.128), in order to establish whether a policy of more active support for straw-processing within the Community might be justified. The Panel recommended that the Commission should initiate such a critical study.

Expansion of the European straw-processing industry would lead to the production of increased quantities of hemicellulose and lignin. The Panel considered that more research would then be needed on lignin copolymers, grafting and blends, and on the utilisation of hemicellulose derivatives to obtain in one step, from furfural, derivatives which can be used in polymer chemistry and other areas including the specific properties of the furanic heterocycle. However, increased effort in these subjects will only be justified if there is likely to be increased use of straw-pulp in paper, as this will continue to be the primary enterprise.

1.5 General

The Panel recognised that present industrial use of surplus carbohydrates in the Community is restricted by their high price, and welcomed the current Commission policy of bringing prices closer to 'world' prices. The Panel considered however that a further impediment to greater usage by industry is that some potential industrial users lack confidence in the ability of European agriculture to guarantee either security of supply or consistent quality in raw materials; this seems particularly the case with cereal straw. The Panel thus considered that serious attention should be given to the potential role of the Agricultural Refinery (Rexen and Munck, pp.190-214) in (a) providing a 'buffer' between field production and the supply to industry of a range of raw materials: (b) carrying out limited local processing and concentration before raw materials are transferred to the industrial users: and (c) reducing the need for high capital investment in machinery on small and medium sized farms, so improving their viability (of particular social importance in the Community).

The Panel recommended that the Commission should set in hand a critical and independent assessment of the Agricultural Refinery concept, to determine in particular whether it would be likely, in practice, to increase the usage by industry of agricultural raw materials.

DISCUSSION PANEL 2 : FUEL PRODUCTS AND ADDITIVES

Chairman: D. SCHLIEPHAKE (BRD)

2.1 Technical and Economic Background

The Panel mainly discussed the role of ethanol, which is the main potential energy product of agriculture.

2.1.1 Ethanol as a cosolvent

The use of methanol, a cheap industrial energy source, as an adjuvant to gasoline shows, in the presence of water, low solubility in hydrocarbons. Solubility can be increased if methanol is mixed with the less polar ethanol: possible additions would be 3% V/V methanol plus 2-3% V/V anhydrous ethanol with gasoline.

The main competitor for this purpose is tertiary butanol (TBA). This is cheaper than ethanol, but current industrial production is only 900 000 tonnes per year (1.15×10^6 m^3). Gasoline consumption in the EEC in 1984 was 85×10^6 tonnes (113×10^6 m^3). Thus a 2% addition level (with 3% methanol) would require all the available TBA plus 1.11×10^6 m^3 ethanol; at the 3% level ethanol requirement would increase to 2.24×10^6 m^3.

2.1.2 Ethanol as an additive (octane enhancer)

In order to decrease the consumption of lead in fuels various combinations of ethanol, with or without TBA and methanol, can be envisaged:

10% ethanol (V/V), according to EC directive only 5% (hydrated)

or 2% TBA/3% methanol/5% ethanol, (hydrated or anhydrous)

or 3% methanol/5% ethanol (anhydrous)

These or other combinations will depend on the octane demand and the quality of the basic gasoline: critical properties are the Motor Octane Number, Road Octane Number and Vapour Pressure. The main competitors to ethanol will be the methyl ether of tertiary butyric acid (MTBE) which gives better performances than ethanol, methanol plus TBA (oxinol). The market potential for ethanol as an additive can be estimated in different ways:

2.1.2.1 Addition of 10% V/V (maximum) of additive to premium gasoline

(75% of total gasoline = 84.8×10^6 m^3)

This would require 8.48×10^6 m^3 of additive. Ethanol would only be needed after all the available MTBE (currently 1.62×10^6 m^3) and oxinol ($2.88 \times 0.75 \times 10^6$ m^3) is used; thus the maximum market would be 4.70×10^6 m^3.

2.1.2.2 Addition of 5% (V/V) of additive to premium gasoline

(+ 2% TBA and 3% methanol)

At current premium gasoline consumption this would require 4.25×10^6 m^3 of additive. Again all the available MTBE and oxinol would be used first (oxinol having still been added), leaving a balance of 2.62×10^6 m^3 of ethanol required.

Thus under the condition of compulsory lead-free gasoline, the requirement of ethanol could be between 3 and 4×10^6 m^3 per year, with a price close to the premium gasoline price.

2.1.3 Ethanol as a gasoline extender (10% to 85% V/V)

For ethanol to be used at these higher levels present car engines would have to be considerably modified; ethanol would also have to be produced at below current prices because it would then be competing directly with gasoline on energy cost. Higher usage of ethanol would also disturb the present integrated market use of crude oil (as has already occurred in Brazil) and is unlikely to be actively supported by the oil companies.

2.1.4 Addition of ethanol to diesel fuel

This is technically possible; but to be competitive ethanol would have to be available at a low price because of the need for additional cosolvents and ignition-improvers to give satisfactory performance.

2.1.5 Ethanol as a substitute fuel for gasoline

If ethanol could be produced at a very low price it could then find a market as a direct substitute for gasoline or diesel fuel in specifically modified engines, in direct competition with lead-free gasoline.

The potential roles of ethanol (1.1-1.5) can be summarised schematically on the basis of price relative to the price of premium gasoline at Rotterdam.

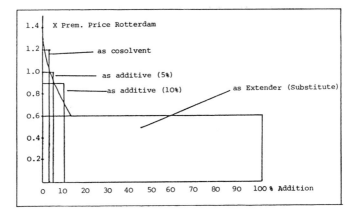

Fig. 1.

Figure 1 shows clearly that, at current relative price (>1.5), there is little probability of a significant usage of ethanol in motor fuels. For this position to change it is suggested that investigation/ research on the following subjects are needed.

2.2 Recommendations

2.2.1

A critical study of the technical requirements for organic additives to gasoline (including ethanol) as the use of lead in gasoline is phased out within the EEC. This study must be carried out in full collaboration with the oil industry, and must take note of specific problems/conditions within the different countries of the Community.

This study should form the basis for political decisions on the rate at which the amount of lead in gasoline is reduced.

2.2.2

A critical study of the technology, infrastructure and economics of the different methods of production of ethanol already used or under development.

2.2.3

Improvement of the genetics, cultivation and harvesting of agricultural crops for ethanol production. This should concentrate in particular on improving the suitability, rather than the yield, of different crops for industrial processing, and also on the composition and value of the byproducts, which can make an important contribution to the overall economics of the process.

2.2.4

Further research and development are needed to improve the efficiency of all stages in the process of producing alcohol from agricultural materials, including handling and preparation, development of continuous fermentation processes and reduction in energy-use in the process. Much of this R&D will be carried out by commercial companies and any work sponsored by the Commission must be fully coordinated with this commercial research.

2.2.5

Finally, the use of ethanol as motor fuel will depend greatly on fiscal policy: a differential taxation vis-a-vis gasoline may be necessary to stimulate usage of ethanol.

2.3 Conclusions

The Panel concluded that there is scope for some use of ethanol in motor fuel (up to perhaps $7 \times 10^6 \text{ m}^3$ per year) as a cosolvent or as an additive to gasoline up to 10% V/V. At present prices, however, ethanol cannot compete as a substitute fuel: for this, either the cost of production must be considerably reduced (which will require further R&D) or its market price vis-a-vis gasoline must be made more favourable with the aid of public funding, e.g. by changing tax regulations.

DISCUSSION PANEL 3 : CHEMISTRY OF PROTEINS AND LIPIDS

Chairman: J.L. MAUBOIS (F)

3.1
It was noted that the members of the Panel were all experts in dairy science, and lacked expertise in proteins and oils derived directly from plant materials. The following comments should therefore be read with that qualification.
It was agreed that direct substitution of one protein, or fat, for an existing product is not a solution as it merely transfers the "surplus" problem from one sector to another. Development of new uses and expansion of existing uses are required, and these would mainly be in the non-food sector. For milk or milk components to compete in the non-food sector considerable reductions in prices would be needed.

3.2
Within this remit the following points were noted.

3.2.1
The problem of surplus production continues to increase in severity, both because levels of output continue to increase (though with some check in the dairy sector following the introduction of milk quotas) and because of continuing competition from alternative products, in particular those derived from vegetable oils (competing with milk fat) and from mycoproteins (competing with milk and meat proteins). Genetic engineering is likely to lead to increased competition of plant with animal products.

3.2.2
There is also likely to be more consumer discrimination against milk (and meat) fat as a result of current concern about possible health hazards from consumption of fat.

3.2.3
Energy costs are likely to fall in real terms (in particular from nuclear reactors) and this will tend to make chemically-synthesised food ingredients cheaper - and so more competitive.

3.2.4
The Panel noted that the Community has a surplus of butter (currently about 900 000 tonnes) but a deficit of protein and vegetable oils. EC support measures are encouraging the production of plant protein and oils, with the aim in particular of reducing the Community's present annual import of 18 million tonnes of soya meal.

3.2.5
The most important factor leading to increased or new uses will be a reduction in the unit price of commodities in surplus.

3.2.6
There is evidence, e.g. in France, that producers are beginning to diversify away from animal towards plant production.

3.3

These observations indicated the need for increased research and development within the following subject areas (no attempt was made to allocate these in order of priority).

3.3.1 Exports

Exports of commodities in surplus avoid the problem of 'substitution' noted in 1. However, it is important that exports are closely tailored to customer's specific requirements: this is now often not the case, and more research is needed on basic and technological factors determining product suitability for specific applications. In particular, research is needed to improve the ease of reconstitution of a wide range of milkpowders, to allow milk and powder colour to be manipulated, to produce milk basis more closely matched to natural food preferences, and to improve control of the production and properties of caseinates. The point was noted that the present system of basing restitution payments on dry matter rather than protein content reduces the incentive to export surplus dairy products.

3.3.2 Food industry

The EC industry is actively seeking to increase the use of milk components in a wide range of foods, with particular importance given to improving the functionality of both proteins and lipids. Progress has been more rapid than in the corresponding research with plant proteins because the more extensive knowledge of the molecular structure of milk proteins has made them well-suited for studies of structure/function relationships, for example by relating chemically- and enzymically-induced structural changes to changes in functionality.

Many plant proteins suffer from the further disadvantage that they are associated with toxic or anti-nutritional factors which must be removed before they can be used in food manufacture: non-protein nitrogen constituents are also often erroneously analysed as protein.

R&D requirements identified include:

3.3.2.1

Further basic knowledge is needed on the composition of milk fat, and of enzyme systems which will allow milk fat to be modified by fractionation, esterification, hydrogenation, etc.

3.3.2.2

Research on the molecular basis of the food functionality of proteins, so that their use by the food industry can be extended through chemical and enzymic modification. Particular attention should be given to new techniques in genetic engineering.

3.3.2.3

Research on the interactions between food components during different processes of food manufacture, in particular in relation to reactions involving milk components. It will be important to include examination of health and nutritional consequences in all these studies.

3.4 Chemical Industry

The Panel recognised the potential importance of using surplus agricultural commodities, including fats and proteins, in industrial processes but was able to give only limited advice in this sector. It considered that a critical study of this subject, with full comparison of competitive materials and costs, was urgently needed. The importance of relative prices of raw materials was clear: a rise in the price of hydrocarbons, relative to milk or vegetable fats, could open up important new markets.

This would be further stimulated by research to improve the quality and to reduce the production costs of existing crop species, and their use in combined animal feed/chemical industries. This could include research to improve the amino-acid composition of plant proteins and the fatty-acid composition of fats to match the potential uses; in particular there is need to develop plants, adapted to European conditions, with high contents of the intermediate-length fatty acids.

3.5 Dietetic and Pharmaceutical Uses. Human Health

The Panel strongly recommend that a specialist group, under the chairmanship of Prof. Maubois, should be convened to consider these aspects of the use of fats and oils, particularly in relation to their use by the chemical industry.

3.6 Reducing the Costs of Production of Milk and Milk Components

The Panel recognised that sale of high value products, e.g. pharmaceuticals, could help to offset the lower returns on fractions, such as butterfat, so as to improve the overall economics of milk production. They concluded however that further research to reduce the cost of milk production is urgently needed, for example by reducing the amounts of concentrates, including imported oil-seeds, fed to dairy cows. In relation to the probable future market they also gave high priority to research aimed at reducing the content of fat in milk and increasing that of protein, and of optimising the composition of both components.

DISCUSSION PANEL 4 : ALTERNATIVE LAND USE

Chairman: V. BIANCARDI (I)

4.1
The Panel concluded that, before considering possible alternative uses for land within the EC, it was important to understand the reasons for the current production of agricultural surpluses, which had led to the search for alternative uses: it might in practice be more logical to seek to change the policies that led to surplus production than to deal with the consequences of surplus production.

4.2
The Panel concluded that this may require some shift in EC support away from primary production towards secondary and tertiary (processing) activities, and from high-income to low-income farmers (interestingly, these are among the major proposals in the Green Discussion Paper on the future of the CAP, issued by DG VI in September 1985).

4.3
The Panel was concerned at the general lack of awareness of the likely future impact of informatics/robotics on the agricultural industry, which they considered likely to be felt well before the impact of biotechnology (genetic engineering) to which most attention is given. They concluded that alternative future uses for labour in rural areas will be at least as important as alternative uses for land - hence the need for new agriculture-based production industries, as well as expansion of service industries in rural areas.

4.4
It is essential that markets for the products from these new industries are identified before they are started: agricultural subsidies ought not just be replaced by industrial subsidies. Development should be market-driven.

4.5
Again in relation to surplus production, the problem might be reduced if more attention were given to quality rather than quantity of production: the EC could accept lower yield in a protein crop if its amino-acid composition was better than that of soyabean. Certainly research should emphasise reduction in costs of production rather than increase in yield (the two are not synonymous): a decrease in the levels of fertiliser and pesticides from those currently used would also have environmental advantage in reducing pollution risks.

4.6
Thus a major thrust in the Panel's conclusions was that the EC priority should be not to produce surplus agricultural products in the first place. Where surpluses are produced, ways should be developed of processing them into marketable products at the small-scale farm level. The Panel was more doubtful about the scope for new large-scale industrial processing of agricultural products, in particular fermentation to produce alcohol, which they considered is unlikely to be competitive with fossil fuels, including

LPG from the Middle East. However there may be a role for farm-scale fermentation processes based on farm 'waste' products - though it is not clear that this will compete with methane production from anaerobic fermentation.

4.7
However the Panel recognised that such measures may <u>not</u> be sufficient to deal with the problem of surplus production, and that it may finally be necessary to seek non-agricultural uses for some land. Among these they considered in particular (a) allocation of more land for recreation (though probably requiring continuing low-level farming activity) and (b) diversion of present crop-land to forestry, because there will be a continuing deficit of timber within the EC. The Panel noted however that little work has been done on genetic improvements of trees, and that this must limit the efficiency of timber production: they advised that application of new genetic techniques could be of particular value when applied to species with a long generation-time.

4.8
Finally the Panel stressed again their view that the potential impact of the new informatics has been underestimated: they considered that, by making information on new techniques rapidly available to farmers throughout the EC, they will speed up the rate of technical change.
This impact should be fully exploited to deal with the problem of surplus production (a) by showing farmers how to reduce their output <u>without</u> reducing their incomes, by adopting inherently more efficient systems of production, and (b) by introducing small-scale on-farm processes for adding value to agricultural products. These measures are likely to be at least as important as seeking alternative uses for agricultural land.

DISCUSSION PANEL 5 : NEW CROPS FOR THE FUTURE

Chairman: C. EERKENS (NL)

The Panel examined the potential and problems in the development and adoption of new crops to replace crop products which are currently in surplus.

5.1

They noted that, independently of market requirements, new crops may be needed so as to reverse the present trend towards crop monoculture (particularly with cereals), which carries long term risks to good soil conservation, crop health, etc.

5.2

They also noted that much of the present surplus production results from policy decisions, e.g. on price levels and intervention. These may need to be changed, and specific new incentives to farmers and processors introduced, if 'new' crops are to be introduced successfully (as has already been done with oil-seed crops). Such new actions will also need to be fully coordinated with regional and social policies.

5.3

Research should aim to develop a composite evaluation of crop species in accordance to data on intrinsic production efficiency e.g. plants' energy investments, use and distribution of assimilate to produce useful and unuseful quantities of chemical constituents, including carbohydrates, proteins, lipids, acids, lignins, etc. Furthermore it includes a data evaluation of the energy needed to modify and utilise these constituents in industrial processes. This will require more systematic data collection and interpretation, and this should be applied to the range of materials held in plant collections and gene banks in different climatic zones. Availability of adequate data could greatly speed the adoption of new crops.

5.4

Such data would be of particular importance where the whole crop is harvested and separated into different fractions (the Agricultural Refinery, see 1.5), with the fractions being used in a range of further processes.
The Panel considered that this concept is of importance in relation to potential new crops, because it greatly widens the range of technical options. They supported further research on whole-crop harvesting.

5.5

The Panel confirmed the importance of continued research in particular on animal-feed protein crops, because of the likely continuing production deficit of feed proteins in the EC. They considered however that more attention should be given to non-feed crops, including cuphea, chufa, and oil and fibre crops. They noted the importance of establishing early dialogue with potential users in the processing industries, because user 'pull' is likely to be an essential component in the successful establishment of a new crop.

5.6

    At present prices of fossil fuels, few agricultural crops (except possibly by-products) are likely to be competitive as direct sources of energy. However it is essential to maintain some research effort in this area (e.g. with cultivation of poplar for energy production) as an insurance against possible future increases in energy prices.

5.7

    The Panel concluded that the time, and the range of expertise, available had not been sufficient to allow adequate discussion of this important subject. They recommended that the Commission should organise further detailed discussion within a few months.

A SYNTHESIS OF THE PROPOSALS
IN SCIENCE, TECHNOLOGY AND DEVELOPMENT
RELEVANT TO THE PROBLEM OF AGRICULTURAL SURPLUSES

6.1 Improvement in Agricultural Production

6.1.1 General

Introduction of more efficient systems of production which should permit farmers to reduce output without reducing their income, e.g. by the development of small-scale on-farm processing of agricultural products.

6.1.2 Plant culture

6.1.2.1

Improvement in European plant proteins (amino-acids profile) and fats (fatty acids profile) for nutritional or industrial purposes (including in particular crops producing intermediate-length fatty acids).

6.1.2.2

Data banks on crop characteristics, on the basis of energy stored in carbohydrates, protein quantity and quality, lipids, lignin, organic acids; efficiency of producing these constituents (with emphasis on non-food crop products, like fibres, oils, etc.).

6.1.2.3

Gene banks of crops, with descriptions as in 5.3.

6.1.2.4

Increased productivity of crops to be used for ethanol production.

6.1.2.5

Genetic improvement of trees, using new genetic techniques.

6.1.3 Animal production

6.1.3.1

Reduced cost of milk production, (in particular by reducing the amount of concentrates fed to cows).

6.1.3.2

Production of milk with increased protein content and decreased fat content.

6.2 Processing of Biological Products

6.2.1 Bioreactors [1]

6.2.1.1

The biotransformation of lactose into useful substances.

---

[1] This sector is partly covered by the Biotechnology Action Programme (BAP of the CEC, DG XII).

6.2.1.2
    Downstream processing (isolation, purification, concentration..) of sugar monomers (in polysaccharide hydrolysates) and of hemicellulose derivatives[2]. The saccharides as chiral compounds for the synthesis of novel substances e.g. polymers.

6.2.1.3
    Improvement of ethanol production (fermentation processes, downstream processing).

6.2.1.4
    Improvement of selective reaction methods for the production of specific intermediate chemicals.

6.2.2 Food technology

6.2.2.1
    Reactions between food components.

6.2.2.2
    Development of new human food products of high added value for social groups with particular nutritional requirements (young and aged, athletes, hospitalised and convalescent, sufferers from dietary problems).

6.2.2.3
    Studies on starch: comparison of the starches extracted from different crops grown in the Community; comparison of the starches produced from different wheats (in particular amylose and amylopectin).

6.2.2.4
    Studies on butterfat (fractionation, esterification, hydrogenation, etc.).

6.3 Economical Studies

6.3.1
    Improvement of production schemes for ethanol.

6.3.2
    Support to integrated projects: crop culture - new processing industry - research laboratory to help to start newborn agro-industries.

6.3.3
    Analysis of the possible impact of modifications in the Common Agricultural Policy in terms of economic returns to alternative land use.

6.3.4
    Research on integrated farm systems giving rise to less expensive production, whole-crop harvesting and the agricultural refinery.

----
[2] These subjects only if a previous study has shown the economical value of straw processing for paper pulp production.

LIST OF PROPOSED STUDIES

7.1

The concept of the agricultural refinery as defined in the report of Rexen and Munck: a critical assessment, made independently by various national research groups or consultant teams to be followed by a symposium to make an overall evaluation of agricultural refineries.

7.2

Critical economical, technical and ecological evaluation of the production of paper pulp from straw in the existing factories (for instance in Denmark) and of the new pulping processes now under development: an independent assessment made by various national research groups or consultant teams to be followed by a symposium to make an overall evaluation on production of paper pulp from straw.

7.3

Survey of industrial possibilities for fats and proteins, with full comparison of competitive material and costs.

7.4

The needs for oxygenates in petroleum refineries as a result of legislation to reduce lead content of gasoline. An evaluation country by country.

7.5

Critical review of scientific studies relating dairy products to health (positively or negatively); practical implications.

LIST OF PROPOSED MEETINGS

8.1
Seminar on agricultural surpluses and deficits in fats and oils in the EEC: Research needs and implications for the Common Agricultural Policy.

To be chaired by Pr. Maubois.

8.2
Seminar on starch: the present knowledge on the composition, structure, quality, and technology of starch from various origins. Identification and definition of research requirements.

8.3
Two symposia to make overall evaluation of studies by various European research groups on:

8.3.1
The production of paper pulp from straw.

8.3.2
The concept of the agricultural refinery.

8.4
Sectorial meetings on specific new crops.

LIST OF PARTICIPANTS

BANKS, W.
Hannah Research Institute
UK AYR KA6 5HL

BARNOUD, F.
CERMAV
BP 68
F-38402 St Martin d'Hères Cedex

BIANCARDI, V.
Istituto de Economica e Politica
Agraria della Facoltà Agraria
Via Filippa Re, 10
I-40126 Bologna

CATHELINAUD, Y.
O.E.C.D.
2 rue André Pascal
F-75575 Paris Cedex 16

COCHET, N.
Université de Technologie
de Compiègne
Division des Procédés
Biotechnologiques
Département Génie Chimique
BP 233
F-60206 Compiègne Cedex

CONWAY, A.G.
An Foras Taluntais
Economics and Rural Welfare Centre
19 Sandymount Avenue
IRL - Dublin 4

CULLETON, N.
CCE - DG VI
Loi 86 - 5/48
200 rue de la Loi
B-1049 Bruxelles

DAMBROTH, M.
Institut für Pflanzenbau und
Pflanzenzüchtung (FAL)
Bundesallee 50
D-3300 Braunschweig

DECEUNINCK, D.
Amylum NV
Burchstraat 10
B-9300 Aalst

DEHANDTSCHUTTER, J.
CCE - DG VI
Loi 86 - 04/33
200 rue de la Loi
B-1049 Bruxelles

DELHEYE, G.
Amylum NV
Burchstraat 10
B-9300 Aalst

DELMAS, M.
Laboratoire de Chimie Organique
et Agrochimie
Ecole Nationale Supérieure de
Chimie de Toulouse
118 route de Narbonne
F-31077 Toulouse Cedex

DI GIULIO, A.
CCE - DG VI
L 86 - 5/69
200 rue de la Loi
B-1049 Bruxelles

EERKENS, C.
Centrum voor Agrobiologisch
Onderzoek (CABO)
Crop Science Section
PO Box 14
Bornsesteeg, 65
NL-6700 AA Wageningen

EVANS, E.W.
Food Research Institute
Shinfield
UK - Reading RG2 9AT

FAUCONNEAU, G.
INRA
149 rue de Grenelle
F-75341 Paris Cedex 07

FELDMANN, J.
Maizena GmbH
· Dusseldorfer Strasse 191
D-4150 Krefeld-Linn

GABELLIERI, R.
Centre d'Etudes et d'information
sur les Communautés Européennes
135 rue Stévin
B-1040 Bruxelles

GASET, A.
Laboratoire de Chimie Organique
et d'Agrochimie
Ecole Nationale Supérieure
de Chimie de Toulouse
118 route de Narbonne
F-31077 Toulouse Cedex

GENGHINI, M.
Istituto di Economica e Politica
Agraria della Facoltà Agraria
Via Filippa Re 10
I-40126 Bologna

GILLOT, J.
CCE - DG VI
L-86 - 04/43
200 rue de la Loi
B-1049 Bruxelles

HAEBLER, C.
CCE - DG VI
Berl. 05/23
200 rue de la Loi
B-1049 Bruxelles

HEPNER, L.
Managing Director
Hepner Associates Ltd
Tavistock House North
Tavistock Square
UK - London WC1

HIRSINGER, F.
Henkel KGaA
Postfach 1100
D-4000 Düsseldorf

HOPKINS, A.
Courtaulds plc
18 Hanover Square
UK - London W1A 2BB

JONAS, D.
MAFF
R454 Great Westminster House
Horseferry Road
UK - London SW1

LACIRIGNOLA, C.
Centre International des Hautes
Etudes Méditerranéennes
11 rue Newton
F-75116 Paris

LARVOR, P.
CEC
DG XII - SDM 3/73
200 rue de la Loi
B-1049 Bruxelles

MAHY, D.
Faculté Notre Dame de la Paix
Faculté des Sciences Economiques
et Sociales
Rempart de la Vierge 8
B-5000 Namur

MAUBOIS, J.L.
Laboratoire de Technologie Laitière
Ecole Nationale Supérieure
Agronomique
65 rue de Saint Brieuc
F-35042 Rennes Cedex

MEESTER, G.
Agricultural Economics Research
Institute
PO Box 29703
NL-2502 LS Den Haag

MILDON, R.
CEC
DG VI (Agriculture)
Assistant au Directeur Général
Adjoint chargé des Marches
200 rue de la Loi
B-1049 Bruxelles

MOLLE, J.F.
CEMAGREF
Parc de Tourvoie
BP 121
F-92164 Antony Cedex

MOREALE, A.G.
CEC
DG VI - Berl. 5/5
200 rue de la Loi
B-1049 Bruxelles

MORTENSEN, B.K.
The Danish Government
Research Institute for
Dairy Industry
Roskildevej 56
DK-3400 Hilleroed

MORTENSEN, K.
Head of Section
Ministry of Agriculture
Slotsholmsgade 10
DK-1216 Copenhagen K

MULCAHY, M.J.
Assistant Director 3
An Foras Taluntais
Moore Park Research Centre
Fermoy
Co. Cork, IRL

MUNCK, L.
Department of Biotechnology
Carlsberg Research Center
Gamle Carlsbergvej 10
DK-2500 Copenhagen Valby

DE NETTANCOURT, D.
CEC
DG XII - SDM 2/76
200 rue de la Loi
B-1049 Bruxelles

NIMZ, H.H.
Bundesforschungsanstalt für
Forst und Holzwirtschaft
Leuchnerstrasse 91
D-2050 Hamburg 80

PEYSSARD
ACTA
MNE
149 rue de Bercy
F-75595 Paris Cedex 12

PUGLISI, P.P.
Università di Parma
Istituto di Genetica
Borgo Carissimi 10
I-43100 Parma

QUADFLIEG, H.
TÜV Rheinland
Am grauen Stein
D-5000 Köln 91

RAUGEL, P.J.
Consultant of CIRTA
25 rue Pierre et Marie Curie
F-94200 Ivry sur Seine

RAYMOND, W.F.
Consultant
Periwinkle Cottage
Christmas Common
Watlington
UK - Oxon OX9 5HR

REINIGER, P.
CEC
DG XII - SDM 3/65
200 rue de la Loi
B-1049 Bruxelles

REXEN, F.
Department of Biotechnology
Carlsberg Research Center
Gamle Carlsbergvej 10
DK-2500 Copenhagen Valby

RINAUDO, M.
CNRS
Centre d'Etudes et de Recherches
sur les Macromolécules Végétales
BP 68
F-38402 St Martin d'Hères Cedex

RÖRSCH, A.
TNO
Juliana van Stolberglaan 148
Postbus 297
NL-2501 's Gravenhage BD

RYNJA, S.
Managing Director
IBVL
Bornsesteeg 59
PO Box 18
NL-6708 PD Wageningen

SCHLIEPHAKE, D.
An der Zickelburg 17
D-5340 Bad Honnef

SCULLY, J.
CEC
DG VI - L-86
200 rue de la Loi
B-1049 Bruxelles

SKOV LARSEN, C.
Director Biotechnical Institute
Holbergsvej 10
BK-6000 Kolding

SPEDDING, C.R.W.
Dept. of Agriculture and
Horticulture
University of Reading
1 Earley Gate
UK - Reading RG6 2AT

STRINGER, D.A.
ICI plc
Agricultural Division
PO Box 1
Billingham
UK - Cleveland TS23 1LB

VONTHRON, J.R.
CEC
DG VI - Berl. 5/10
200 rue de la Loi
B-1049 Bruxelles

WERRY, P.A.Th.J.
Directorate of Agricultural
Research
PO Box 59
NL-6708 PA Wageningen

WHITE, D.J.
MAFF
Room 120
Chief Scientists Group
Great Westminster House
Horseferry Road
UK - London SW1P 2AE

WOELK, H.U.
Geschäftsführer
Maizena GmbH
Spaldingstrasse 218
D-2000 Hamburg 1